Handbook of Collective Intelligence

Handbook of Collective Intelligence

Thomas W. Malone and Michael S. Bernstein, editors

The MIT Press
Cambridge, Massachusetts
London, England

© 2015 Massachusetts Institute of Technology

All rights reserved. No part of this book may be reproduced in any form by any electronic or mechanical means (including photocopying, recording, or information storage and retrieval) without permission in writing from the publisher.

Set in Palatino by the MIT Press. Printed and bound in the United States of America.

Cataloging-in-Publication information is available from the Library of Congress.

ISBN: 978-0-262-02981-0

10 9 8 7 6 5 4 3 2 1

For Virginia. —T.M.

For Marvin, Ethel, Leonard, and Sylvia. —M.B.

Contents

Acknowledgments

We are grateful to the MIT Center for Collective Intelligence, the MIT Sloan School of Management, and the Stanford Computer Science Department for support during the preparation of this book, and to the National Science Foundation for support of the first two Collective Intelligence Conferences, held in 2012 and 2014, which helped make the field this book describes a reality (NSF grant number IIS-1047567). In addition, Malone's work on this volume was supported in part by grants from the National Science Foundation (grant numbers IIS-0963285, ACI-1322254, and IIS-0963451), and the U.S. Army Research Office (grant numbers 56692-MA and 64079-NS). Bernstein's work on this volume was supported in part by a grant from the National Science Foundation (IIS-1351131).

We are also grateful to Richard Hill for excellent administrative assistance in many stages of this work, to Lisa Jing for bibliographic research used in the introduction, and to the members of the online "crowd" who commented on earlier versions of the chapters.

Introduction

Thomas W. Malone and Michael S. Bernstein

In nine hours, a team successfully scoured the entire United States to find a set of red balloons that was worth $40,000 (Pickard et al. 2011). In three weeks, citizen scientists playing a game uncovered the structure of an enzyme that had eluded scientists for more than fifteen years (Khatib et al. 2011). In ten years, millions of people authored the most expansive encyclopedia in human history. If interconnected people and computers can accomplish these goals in hours, days, and years, what might be possible in the next few years, or the next ten?

This book takes the perspective that intelligence is not just something that arises inside individual brains—it also arises in groups of individuals. We call this *collective intelligence*: groups of individuals acting collectively in ways that seem intelligent (Malone, Laubacher, and Dellarocas 2009). By this definition, collective intelligence has existed for a very long time. Families, armies, countries, and companies have all—at least sometimes—acted collectively in ways that seem intelligent. And researchers in many fields—from economics to political science to psychology—have studied these different forms of collective intelligence.

But in the last two decades a new kind of collective intelligence has emerged: interconnected groups of people and computers, collectively doing intelligent things. For example, Google harvests knowledge generated by millions of people creating and linking Web pages and then uses that knowledge to answer queries in ways that often seem amazingly intelligent. In Wikipedia, thousands of people around the world have collectively created a large intellectual product of high quality with almost no centralized control and with mostly volunteer participants. And in more and more domains, surprisingly large groups of people and computers are writing software (Lakhani, Garvin, and Lonstein 2010; Benkler 2002), solving engineering problems (Lakhani and

Lonstein 2011), composing and editing documents (Kittur, Smus, Khamkar, and Kraut 2011; Bernstein et al. 2010), and predicting presidential elections (Berg, Forsythe, Nelson, and Rietz 2008).

These early examples, we believe, are not the end of the story but just the beginning. And in order to understand the possibilities and constraints of these new kinds of intelligence, we need a new interdisciplinary field. Such a field can help in the exploitation of the often-unrecognized synergies among disciplines that have studied various forms of collective intelligence without realizing their commonalities, and can develop new knowledge that is specifically focused on understanding and creating these new kinds of intelligence. Helping to form such a field is the primary goal of this volume.

Defining Collective Intelligence

As with many important but evocative terms, there have been almost as many definitions of collective intelligence as there have been writers who have described it (see the appendix to this introduction for a representative list). For instance, Hiltz and Turoff (1978) define collective intelligence as "a collective decision capability [that is] at least as good as or better than any single member of the group." Smith (1994) defines it as "a group of human beings [carrying] out a task as if the group, itself, were a coherent, intelligent organism working with one mind, rather than a collection of independent agents." And Levy (1994) defines it as "a form of *universally distributed intelligence*, constantly enhanced, coordinated in real time, and resulting in the effective mobilization of skills."

Of course, intelligence itself can be defined in many different ways. Sometimes it is defined in terms of specific processes—for example, "intelligence is a very general mental capability that, among other things, involves the ability to reason, plan, solve problems, think abstractly, comprehend complex ideas, learn quickly and learn from experience" (Gottfredson 1997). Another common way of defining intelligence is in terms of goals and the environment—for example, in 2006 the *Encyclopaedia Britannica* defined it as "the ability to adapt effectively to the environment," and in 1983 Howard Gardner defined it as "the ability to solve problems, or to create products, that are valued within one or more cultural settings." The most common operational definition of intelligence in psychology is as a statistical factor that measures a person's ability to perform well on a wide range of very different cognitive tasks (Spearman 1904). This factor (often called "general intelligence" or g) is essentially what is measured by IQ tests. There is even controversy about whether it would be legitimate to call behavior intelligent, no matter how intelligent it seemed, if it were produced by a computer rather than a person (see, e.g., Searle 1999).

In view of all this complexity, our definition of collective intelligence, as given above, is a simple one: *groups of individuals acting collectively in ways that seem intelligent*. Several aspects of this definition are noteworthy:

- The definition does not try to define "intelligence." There are so many ways to define it, and we do not want to prematurely constrain what we believe to be an emerging area of study. Our definition is, therefore, compatible with all of the above definitions of intelligence.
- By using the word "acting," the definition requires intelligence to be manifested in some kind of behavior. By this definition, for instance, the knowledge represented in a collection such as Wikipedia would not, in itself, be considered intelligent, but the group of people who created the collection could be.
- The definition requires that, in order to analyze something as collective intelligence, one be able to identify some group of individuals who are involved. In some cases this may be straightforward, such as noting the individual humans in an organization; in other cases, it may be useful to draw these boundaries in unusual ways. For instance, if one regards the whole brain as a group of individual neurons or brain regions, one may analyze the operation of a single human brain as collective intelligence. Or one may analyze the collective intelligence of a whole economy by noting that the economy is a collection of many different organizations and people.
- The definition requires that the individuals act *collectively*—that is, that there be some relationships among their activities. We certainly do not intend this to mean that they must all share the same goals or that they must always cooperate. We merely mean that their activities are not completely independent—that there are some interdependencies among them (see, e.g., Malone and Crowston 1994). For instance, different actors in a market may buy and sell things to one another even though they may each have very different individual goals. And different problem solvers in an open-innovation community such as InnoCentive compete to develop the best solutions to a problem.
- By using the word "seem," the definition makes clear that what is

considered intelligent depends on the perspective of the observer. For instance, to evaluate whether an entity is acting intelligently an observer may have to make assumptions about what the entity's goals are. IQ tests, for example, do not measure intelligence well if the person taking the test is only trying to annoy the person giving the test. Or an observer may choose to analyze how intelligently a group of Twitter users filters information even if none of the individual users have that goal.

How Does Collective Intelligence Relate to Other Fields?

In establishing an interdisciplinary field such as collective intelligence, it is useful to indicate how the new field overlaps with and differs from existing fields. In that spirit, we suggest the following loose guidelines for thinking about how collective intelligence relates to several existing fields.

Computer Science

Collective intelligence overlaps with the subset of computer science that involves intelligent behavior by groups of people, computers, or both. For instance, "groups" of one person and one computer (human–computer interaction) can be viewed as a kind of collective intelligence, and studying larger groups of people and computers (e.g., human computation, crowdsourcing, social software, computer-supported cooperative work, groupware, collaboration technology) is at the heart of the field. Similarly, studies of how *groups* of artificially intelligent agents can exhibit intelligent behavior together can be very relevant for collective intelligence, but studies that *don't* focus on how different processing units work together probably are not.

Cognitive Science

Cognitive science focuses on understanding the nature of the human mind, including many aspects of mental functioning that may be regarded as components of intelligent behavior (such as perception, language, memory, and reasoning). Collective intelligence overlaps with cognitive science only when there is an explicit focus on how intelligent behavior arises from groups of individuals. Most obviously, this occurs with groups of people (as in group memory, group problem solving, and organizational learning), but, as was noted above, studies of how different parts of a single brain interact to produce intelligent behavior can also be part of collective intelligence.

Sociology, Political Science, Economics, Social Psychology, Anthropology, Organization Theory, Law

These fields all study the behavior of groups. They overlap with collective intelligence only when there is a focus on overall collective behavior that can be regarded as more or less intelligent. For instance, analyzing how individual people's attitudes are determined or how they make economic choices would not be central to collective intelligence, but analyzing how different regulatory mechanisms in markets lead to more or less intelligent behavior by the markets as a whole would be central to collective intelligence. Similarly, analyzing how different organizational designs in a company lead to better or worse performance by the company as a whole would also be central to collective intelligence. And so would analyzing how well democratic governments make decisions and solve problems.

Biology

Collective intelligence overlaps with the parts of biology that focus explicitly on group behavior that can be regarded as intelligent. For instance, studies of beehives and ant colonies sometimes focus on how the individual insects interact to produce overall behavior that is adaptive for the group.

Network Science

Collective intelligence focuses on the subset of network science that involves intelligent collective behavior. For instance, simply analyzing how rapidly news diffuses in networks with different topologies would not be central to collective intelligence, because there is no overall intelligent behavior being explicitly analyzed. But if such a study also analyzed how effectively the network as a whole filtered different kinds of news or how the speed of information diffusion affected the speed of problem solving, it would be central to collective intelligence.

The History of Collective Intelligence as a Topic of Study

The phrase "collective intelligence" has been used descriptively since the 1800s if not longer. For instance, the physician Robert Graves (1842, pp. 21–22) used it to describe the accelerating progress of medical knowledge, the political philosopher John Pumroy (1846, p. 25) used it to describe the people's sovereignty in government, and C. W. Shields (1889, pp. 6–7) used it to describe science as a collective endeavor. In 1906, the sociologist Lester Frank Ward used the term in something like its modern sense (Ward 1906, p. 39): "The extent to which [society will evolve] will depend upon the collective intelligence. This is to society what brain power is to the individual."

The earliest scholarly article we have found with "collective intelligence" in the title was by David Wechsler, the psychologist who developed some of the most widely used IQ tests (Wechsler 1971). In it Wechsler argues that collective intelligence is more than just collective behavior; it also involves cross-fertilization resulting in something that could not have been produced by individuals. Around the same time, the computer scientist Doug Engelbart was doing his pioneering work on "augmenting human intellect" with computers, including computational support for team cooperation (Engelbart 1962, p. 105; Engelbart and English 1968). Later Engelbart used the term "collective IQ" to describe such work and its broader implications (e.g., Engelbart 1995).

In 1978, Roxanne Hiltz and Murray Turoff used the term "collective intelligence" to describe the goal of the computerized conferencing systems they pioneered (Hiltz and Turoff 1978). In the 1980s and the 1990s, "collective intelligence" was used more and more to describe the behavior of insects (Franks 1989), of groups of mobile robots (Mataric 1993), of human groups (Pór 1995; Atlee 1999; Isaacs 1999), and of electronically mediated human collaboration (Smith 1994; Levy 1994; Heylighen 1999). As best we can tell, the first two books with "collective intelligence" in their titles also appeared during that period. Smith's (1994) focused on computer-supported work groups, Levy's (1994) on the worldwide exchange of ideas in cyberspace.

In the 2000s, the term "collective intelligence" was used even more widely—in some of the publications mentioned later in this volume and in many other publications in computer science, spirituality, and business (e.g., Szuba 2001; Hamilton 2004; O'Reilly 2005; Segaran 2007; Jenkins 2008; Howe 2009). Of particular importance to the spread of the concept were *The Wisdom of Crowds* (Surowiecki 2004) and other books for a general audience featuring the concept of collective intelligence—e.g., *Wikinomics* (Tapscott and Williams 2006) and *The Rational Optimist* (Ridley 2010).

The 2000s also saw the first academic conferences on collective intelligence (Kowalczyk 2009; Bastiaens, Baumol, and Kramer 2010; Malone and von Ahn 2012) and the first academic research centers focusing specifically on this topic (the Canada Research Chair in Collective Intelligence, started in 2002 at the University of Ottawa; and the Center for Collective Intelligence, started in 2006 at the Massachusetts Institute of Technology).

Related Concepts

In addition to those who have used the specific term "collective intelligence," writers in many fields have talked about closely related concepts. For example, psychologists have talked about similar concepts since the 1800s, including crowd psychology (Tarde 1890), crowd mind (Le Bon 1895; Freud 1922), and the collective unconscious (Jung 1934). Emile Durkheim (1893) used the term "collective consciousness" for the shared beliefs and values that lead to group solidarity. Adam Smith (1795) talked about an "invisible hand" controlling allocation of resources in a market. And several writers talked about forms of collective intelligence on a global scale, using terms such as "world brain" (Wells 1938), "planetary mind" (Teilhard de Chardin 1955), and "global brain" (Russell 1983; Bloom 2000).

More recently, social scientists have discussed numerous phenomena that can be regarded as examples of collective intelligence. For instance, theories of transactive memory analyze how groups (such as married couples and business organizations) divide and coordinate the work of remembering things (Wegner, Giuliano, and Hertel 1985; Wegner 1986). Studies of distributed cognition examine how real-world cognitive processes (such as navigating a naval vessel) do not occur solely within a single brain but instead include other people and objects as critical components (Hutchins 1995a,b). Studies of trust in testimony illuminate how much of what we think we know about science, religion, and other topics is based, not on our own firsthand experience, but on what others we trust have told us (Harris and Koenig 2006). And philosophers have argued that it is arbitrary to say that the mind is contained only within the skull. Instead, many cognitive processes are actively coupled with external objects, such as the pen and paper used to do long multiplication and the rearrangement of letter tiles to prompt word recall in Scrabble, and thus these external objects should be considered part of the "mind" that carries out the cognitive processes (Clark and Chalmers 1998). Some of these concepts, and many others closely related to collective intelligence, will be discussed in later chapters.

Why Is Collective Intelligence a Timely Topic Now?

The past few years have seen a significant increase in the popularity and maturity of research in collective intelligence. As was noted above, new forms of collective intelligence made possible by information technology are affecting the daily lives of great numbers of people all over the world. And many disciplines, including neuroscience, economics, and biology, are making fundamental breakthroughs in understanding how groups of individuals can collectively do intelligent things.

But if we do not make an effort to synthesize these insights across fields, we will end up with silos of knowledge, redundant efforts in different academic communities, and lost opportunities for interdisciplinary synergies. We believe this is a critical time for these different fields to come together and begin sharing insights.

This urgency is balanced by pragmatic considerations: the scope of the challenge is large, so we must draw on all the resources at our disposal to tackle them. Several of the authors in this volume have already crossed disciplinary lines in their research. Our goal is to help others to do the same. We hope to provide readers with the tools they need to know when each disciplinary perspective might be useful and with leads to follow when they want to find out more.

An Overview of the Book

This book aims to help coalesce the field of collective intelligence by laying out a shared set of research challenges and methodological perspectives from a number of different disciplines. In the chapters that follow, authors from each discipline introduce the discipline's foundational work in collective intelligence, including its methods (such as system engineering, controlled experiment, naturalistic observation, mathematical proof, and simulation), its important research questions, and its main results. The chapters and their introductions include references to classic works and to recent research results, providing both places to start to learn more and the beginnings of a shared set of references for this field.

Since our goal in organizing the chapters is to stimulate useful connections, we do not believe there is any one best way of grouping different disciplinary topics. Accordingly, the grouping of disciplines in the chapters that follow is somewhat arbitrary, depending not only on intellectual considerations but also on the availability of potential chapter authors.

We begin with what are, perhaps, the most surprising kinds of collective intelligence. First, Andrew Lo summarizes what the field of economics can teach us about how collective intelligence happens (and sometimes doesn't happen) in markets. Then Deborah Gordon introduces a biological-sciences perspective on collective intelligence.

Next we move on to the newest kinds of collective intelligence: those enabled by computers. In two separate chapters, one on human–computer interaction and one on artificial intelligence, Jeff Bigham, Michael Bernstein, Eytan Adar, Daniel Weld, Mausam, Christopher Lin, and Jonathan Bragg lay out computer science's interest in building platforms for crowdsourcing, guiding complex crowd tasks, automating workflows, quality control, and creating hybrid artificial intelligence / crowd systems.

The last three chapters focus on what the social sciences (other than economics) can teach us about collective intelligence involving people. First, Mark Steyvers and Brent Miller introduce ideas from cognitive psychology that are relevant to one important kind of collective intelligence: the "wisdom of crowds" effect. Then, Anita Williams Woolley, Ishani Aggarwal, and Thomas Malone describe perspectives from social psychology and organizational theory, introducing ideas about how human groups work and how they can be collectively intelligent. Finally, Yochai Benkler, Aaron Shaw, and Benjamin Mako Hill focus on "peer production" as a form of collective intelligence. Though this chapter focuses on one specific theme, the chapter and the editors' introduction represent a much wider range of work that is relevant to collective intelligence in law, communications, sociology, political science, and anthropology

At first glance, these disciplines and their writings may seem disconnected, a series of ivory towers with no bridges to connect them. However, in these chapters, the disciplines often focus on the same basic issues using different methodologies and perspectives. For example, many of the chapters grapple with questions such as these:

• What basic processes are involved in intelligent behavior of any kind, collective or not?
• How can these processes be performed by groups of individuals? For instance, how can the processes be divided into separate activities done by different individuals? And what additional activities are needed to coordinate the separate activities?
• What kinds of incentives and other design elements can lead to coherent behavior from the overall group?

Pairs of chapters also complement one another. For example, the chapter on biology and the chapter on artificial intelligence both explore mechanisms for deciding how many members of a collective to place on a task. Similarly, the chapter on human–computer interaction views Wikipedia as a set of design decisions that can inspire future systems while the chapter on peer production studies the design of Wikipedia at a micro level to understand its success.

Together, we hope, these foundations will help researchers from many disciplines to join together in tackling what we believe are some of the most exciting research challenges of our age: How can groups of individuals—collectively—be more intelligent than any of their members? How can new combinations of people and computers be more intelligent than any person, group, or computer has ever been before?

Appendix: Representative Definitions of Collective Intelligence

a collective decision capability [that is] at least as good as or better than any single member of the group (Hiltz and Turoff 1978)

a form of universally distributed intelligence, constantly enhanced, coordinated in real time, and resulting in the effective mobilization of skills (Levy 1994)

a group of human beings [carrying] out a task as if the group, itself, were a coherent, intelligent organism working with one mind, rather than a collection of independent agents (Smith 1994)

the ability of a group to "find more or better solutions than … would be found by its members working individually" (Heylighen 1999)

the *intelligence* of a *collective*, which arises from one or more *sources* (Atlee 2003)

the general ability of a group to perform a wide variety of tasks (Woolley et al. 2010)

harnessing the power of a large number of people to solve a difficult problem as a group [which] can solve problems efficiently and offer greater insight and a better answer than any one individual could provide (*Financial Times* Lexicon 2013)

the capacity of biological, social, and cognitive systems to evolve toward higher order complexity and harmony (Pór 2004)

a type of shared or group intelligence that emerges from the collaboration and competition of many individuals and appears in consensus decision making in bacteria, animals, and computer networks (Wikipedia 2013)

References

Atlee, T. 1999. Co-intelligence and community self-organization. *The Permaculture Activist*, December 1999: 4–10 (http://www.co-intelligence.org/CIPol_permacultureCI.html).

Atlee, T. 2003. Defining "Collective Intelligence" The Co-Intelligence Institute (http://co-intelligence.org/CollectiveIntelligence2.html#definitions). Retrieved March 2, 2013.

Bastiaens, T. J., U. Baumol, and B. J. Kramer, eds. 2010. *On Collective Intelligence*. Springer.

Benkler, Y. 2002. Coase's penguin, or, Linux and the nature of the firm. *Yale Law Journal* 112 (3): 369–446.

Berg, J., R. Forsythe, F. Nelson, and T. Rietz. 2008. Results from a dozen years of election futures markets research. In *Handbook of Experimental Economics Results*, ed. C. R. Plott and V. L. Smith. Elsevier.

Bernstein, M. S., G. Little, R. C. Miller, B. Hartmann, M. S. Ackerman, D. R. Karger, D. Crowell, and K. Panovich. 2010. Soylent: A word processor with a crowd inside. In *Proceedings of the 23nd Annual ACM Symposium on User Interface Software and Technology*. ACM.

Bloom, Howard. 2000. *Global Brain: The Evolution of Mass Mind from the Big Bang to the 21st Century*. Wiley.

Clark, A., and D. J. Chalmers. 1998. The extended mind. *Analysis* 58 (1): 10–23.

Durkheim, E. 1893. *De la division du travail social*. English edition: *The Division of Labor in Society* (Free Press, 1984).

Engelbart, D. C. 1962. Augmenting Human Intellect: A Conceptual Framework. Summary Report AFOSR-3223, Stanford Research Institute.

Engelbart, D. C. 1995. Toward augmenting the human intellect and boosting collective IQ. *Communications of the ACM* 38 (8): 30–32. doi:10.1145/208344.208352.

Engelbart, D. C., and W. K. English. 1968. A research center for augmenting human intellect. In AFIPS Conference Proceedings of the 1968 Fall Joint Computer Conference, San Francisco.

Financial Times Lexicon. 2013. Collective intelligence (http://lexicon.ft.com/Term?term=collective-intelligence).

Franks, N. R. 1989. Army ants: A collective intelligence. *American Scientist* 77 (2): 138–145.

Freud, S. 1922. *Group Psychology and the Analysis of the Ego*, tr. J. Strachey. Boni and Liveright.

Gardner, H. 1983. *Frames of Mind: Theory of Multiple Intelligences*. Basic Books.

Gottfredson, L. S. 1997. Mainstream science on intelligence: An editorial with 52 signatories, history, and bibliography. *Intelligence* 24 (1): 13–23.

Graves, R. 1842. *Clinical Lectures*. Barrington & Haswell.

Hamilton, C. 2004. Come together: Can we discover a depth of wisdom far beyond what is available to individuals alone? *What Is Enlightenment* May–July 2004 (http://www.enlightennext.org/magazine/j25/collective.asp).

Harris, P. L., and M. A. Koenig. 2006. Trust in testimony: How children learn about science and religion. *Child Development* 77 (3): 505–524.

Heylighen, F. 1999. Collective intelligence and its implementation on the Web: Algorithms to develop a collective mental map. *Computational and Mathematical Organization Theory* 5 (3): 253–280. doi: 10.1023/A:1009690407292 10.1023/A:1009690407292

Hiltz, S. R., and M. Turoff. 1978. *The Network Nation: Human Communication via Computer.* Addison-Wesley.

Howe, J. 2009. *Crowdsourcing: Why the Power of the Crowd Is Driving the Future of Business.* Crown Business.

Hutchins, E. 1995a. *Cognition in the Wild.* MIT Press.

Hutchins, E. 1995b. How a cockpit remembers its speeds. *Cognitive Science* 19: 265–288.

Isaacs, W. 1999. *Dialogue and the Art of Thinking Together.* Doubleday Currency.

Jenkins, H. 2008. *Convergence Culture: Where Old and New Media Collide.* NYU Press.

Jung, C. G. 1934. *The Archetypes of the Collective Unconscious* (1981 second edition, *Collected Works*, Bollingen), volume 9, part 1.

Kittur, A., B. Smus, S. Khamkar, and R. E. Kraut. 2011. Crowdforge: Crowdsourcing complex work. In *Proceedings of the 24th Annual ACM Symposium on User Interface Software and Technology.* ACM.

Khatib, F., F. Dimaio, S. Cooper, M. Kazmierczyk, M. Gilski, S. Krzywda, H. Zabranska, et al. 2011. Crystal structure of a monomeric retroviral protease solved by protein folding game players. *Nature Structural & Molecular Biology* 18 (10): 1175–1177.

Kowalczyk, R., ed. 2009. *Computational Collective Intelligence.* Springer.

Lakhani, K. R., D. A. Garvin, and E. Lonstein. 2010. TopCoder (A): Developing Software through Crowdsourcing. Harvard Business School General Management Unit Case No. 610–032. Available at http://ssrn.com/abstract=2002884.

Lakhani, K. R., and E. Lonstein. 2011. InnoCentive.Com (B). Harvard Business School General Management Unit Case No. 612–026. Available at http://ssrn.com/abstract=2014058.

Le Bon, G. 1895. *Psychologie des Foules.* 1963 edition: Presses Universitaires de France.

Levy, P. 1994. *L'intelligence collective: Pour une anthropologie du cyberspace.* Editions La Decouverte. English edition: *Collective Intelligence: Mankind's Emerging World in Cyberspace* (Plenum, 1997).

Malone, T. W., and K. Crowston. 1994. The interdisciplinary study of coordination. *ACM Computing Surveys* 26 (1): 87–119.

Malone, T. W., R. Laubacher, and C. N. Dellarocas. 2009. Harnessing Crowds: Mapping the Genome of Collective Intelligence. MIT Center for Collective Intelligence Working Paper; MIT Sloan School of Management Research Paper No. 4732–09.

Malone, T. W., and L. von Ahn, chairs. 2012. Proceedings of the Collective Intelligence 2012. http://arxiv.org/abs/1204.2991v3.

Mataric, M. 1993. Designing emergent behaviors: From local interactions to collective intelligence. In *Proceedings of the Second International Conference on Simulation of Adaptive Behavior.* MIT Press.

O'Reilly, T. 2005. What Is Web 2.0? Design Patterns and Business Models for the Next Generation of Software. Technical Report, O'Reilly Media (http://oreilly.com/web2/archive/what-is-web-20.html).

Pickard, G., et al. 2011. Time-critical social mobilization. *Science* 334: 509–512.

Pór, G. 1995. The quest for collective intelligence. In *Community Building: Renewing Spirit and Learning in Business*, ed. K. Gozdz. New Leaders Press.

Pór, G. 2004. About. Blog of Collective Intelligence (http://blogofcollectiveintelligence.com/about/). Retrieved March 2, 2013.

Pumroy, J. 1846. *The Annual Address Delivered before the Diagnothian and Goethean Literary Societies of Marshall College, Mercersburg, Pa, On the Connection between Government, and Science and Literature*. McKinley and Lescure.

Ridley, M. 2010. *The Rational Optimist: How Prosperity Evolves*. HarperCollins.

Russell, P. 1983. *The Global Brain: Speculations on the Evolutionary Leap to Planetary Consciousness*. Tarcher.

Searle, J. 1999. The Chinese room. In *The MIT Encyclopedia of the Cognitive Sciences*, ed. R. A. Wilson and F. Keil. MIT Press.

Segaran, T. 2007. *Programming Collective Intelligence*. O'Reilly Media.

Shields, C. W. 1889. *Philosophia Ultima*, volume II. Scribner.

Smith, A. 1795. *The Theory of Moral Sentiments*. Printed for A. Millar in the Strand and A. Kincaid and J. Bell in Edinburgh.

Smith, J. 1994. *Collective Intelligence in Computer-Based Collaboration*. Erlbaum.

Spearman, C. 1904. "General intelligence" objectively determined and measured. *American Journal of Psychology* 15: 201–293.

Surowiecki, J. 2004. *The Wisdom of Crowds*. Random House.

Szuba, T. 2001. *Computational Collective Intelligence*. Wiley.

Tapscott, D., and A. D. Williams. 2006. *Wikinomics: How Mass Collaboration Changes Everything*. Portfolio/Penguin.

Tarde, G. 1890. *Les Lois de l'Imitation*. Ancienne Librairie Germer Baillière.

Teilhard de Chardin, P. 1955. *Le Phénomène Humain*. Editions du Seuil. English edition: *The Phenomenon of Man* (Harper & Row, 1959).

Ward, L. F. 1906. *Applied Sociology*. Ginn.

Wechsler, D. 1971. Concept of collective intelligence. *American Psychologist* 26 (10): 904–907.

Wegner, D. M., T. Giuliano, and P. Hertel. 1985. Cognitive interdependence in close relationships. In *Compatible and Incompatible Relationships*, ed. W. J. Ickes. Springer.

Wegner, D. M. 1986. Transactive memory: A contemporary analysis of the group mind. In *Theories of Group Behavior*, ed. B. Mullen and G. R. Goethals. Springer.

Wells, H. G. 1938. *World Brain*. Methuen.

Wikipedia. 2013. Collective intelligence. http://en.wikipedia.org/wiki/Collective _intelligence. Retrieved March 2, 2013.

Woolley, A. W., C. F. Chabris, A. Pentland, N. Hashmi, and T. W. Malone. 2010. Evidence for a collective intelligence factor in the performance of human groups. *Science 330* (6004): 686–688.

Economics

Editors' Introduction

Markets are among the most important—but, in a certain sense, invisible—examples of collective intelligence in the world around us. When large numbers of buyers and sellers interact, all trying to maximize their own individual self-interest, the "invisible hand" of the market (Smith 1776) allocates the resources being exchanged in a way that often seems surprisingly intelligent overall. Scarce resources go to the buyers who value them the most. Plentiful resources are distributed to all who want them. And all participants in the market have incentives to produce the things that are most valuable to others.

Of course, the real story is more complicated, and Andrew Lo's chapter provides a fascinating overview of what the field of economics has to teach us about how markets work—and sometimes don't. Drawing on examples that include the financial crisis of 2007–2009 and the stock market's reaction to the 1986 *Challenger* disaster, Lo analyzes the strengths and weaknesses of markets as a kind of collective intelligence mechanism for sharing information and making decisions about how to allocate resources.

Lo also suggests how some of the aspects of market behavior that are most puzzling to traditional economic theorists can be explained by drawing on insights about collective intelligence in another field: the process of evolution as studied by biologists.

In addition to studying how markets work, the field of economics has investigated a number of other topics that can help us to understand and invent collective intelligence in other situations. Much of that work focuses on developing detailed mathematical models of how incentives affect the behavior of groups of rational decision makers. The subfield of game theory, for example, analyzes mathematically how the incentives for individual actors affect conflict and cooperation among individuals (see, e.g., von Neumann and Morgenstern 1944; Myerson 1991; Dixit and Nalebuff 2010). For instance, games in which all the incentives sum to zero (that is, in which one person's gain is always someone else's loss) are different in important ways from non-zero-sum games (such as the "prisoners' dilemma") where some choices can be better or worse for all the players collectively.

There is also an extensive literature in economics on how incentives, property rights, and transaction costs shed light on when hierarchical organizations are a more desirable way than markets of coordinating the activities of actors—see, e.g., Hayek 1945, Coase 1937, and Williamson 1975. For instance, Coase pointed out that businesses have an incentive to perform activities themselves, rather than contract them out to someone else in a market, only when the transaction costs of coordinating the activities internally are less than those of having them done externally.

Other work focuses on how understanding these factors can help design effective organizations—both hierarchies and other kinds of organizations (see, e.g., Milgrom and Roberts 1992; Hurwicz and Reiter 2006; Gibbons and Roberts 2012). For instance, one problem in ordinary auctions is that bidders may understate the true value of an item being sold if they think doing so will give them a strategic bidding advantage. But a special kind of sealed-bid auction, called a Vickrey auction, overcomes this problem by giving the item to the highest bidder while charging that bidder only the amount of the second-highest bid. With this arrangement, bidders are incentivized to submit bids that honestly state the true maximum value they would pay for the item (Vickrey 1961). These questions of designing effective organizations will reappear in greater focus in this volume's chapter on organizational behavior.

Economists have even analyzed decision-making mechanisms such as voting (Condorcet 1785; Arrow 1951). For instance, Condorcet's jury theorem shows that, in general, voting leads to good outcomes only if the average voter is more likely than not to vote for the best decision. And Condorcet also proposed a rank-ordered voting method that elects the candidate who would win by majority rule in all pairings with other candidates, if such a candidate exists. The chapter in this volume on artificial intelligence takes up questions of how to combine votes efficiently, and the chapter on law and other disciplines explores how large collectives such as Wikipedia rely on this kind of decision-making behavior to operate in a distributed fashion.

Much recent work in economics has focused on behavioral economics, the empirical study of how actual human decision making departs from the ideal of completely rational decision making that is assumed by most traditional economic models (see, e.g., Kahneman and Tversky 1979; Camerer, Loewenstein, and Rabin 2003; Brennan and Lo 2011, 2012; Lo 2012). For instance, Kahneman and Tversky found that, from the point of view of mathematical decision theory, people who seek to avoid risk when they have a chance of gaining something are irrationally willing to accept risk when they have a chance of losing. Brennan and Lo explore the evolutionary underpinnings of these and other apparent irrationalities and show that such anomalies often result from natural selection in stochastic environments, conferring survival benefits on certain subpopulations and providing one explanation for the very origin of intelligence. These insights into actual human behavior help to explain a number of ways in which real markets depart from the completely rational ideal of previous theories.

In short, the work described in the chapter that follows—and other parts of economics—can help us to analyze and design collectively intelligent ways to allocate resources and make decisions in many other kinds of systems.

References

Arrow, Kenneth. 1951. *Social Choice and Individual Values*. Yale University Press.

Brennan, Thomas J., and Andrew W. Lo. 2011. The origin of behavior. *Quarterly Journal of Finance* 1: 55–108.

Brennan, Thomas J., and Andrew W. Lo. 2012. An evolutionary model of bounded rationality and intelligence. *PLoS ONE* 7: e50310. doi:10.1371/journal.pone.0050310.

Camerer, Colin F., George Lowenstein, and Matthew Rabin, eds. 2003. *Advances in Behavioral Economics*. Princeton University Press.

Coase, Ronald. 1937. The nature of the firm. *Economica* 4 (16): 386–405.

Condorcet, M. 1785. Essay on the application of analysis to the probability of majority decisions. Reprinted in *Condorcet: Selected Writings*, ed. K. M. Baker. Bobbs-Merrill, 1976.

Dixit, Avinash K., and Barry J. Nalebuff. 2010. *The Art of Strategy: A Game Theorist's Guide to Success in Business and Life*. Norton.

Gibbons, Robert, and John Roberts, eds. 2012. *The Handbook of Organizational Economics*. Princeton University Press.

Hayek, Friedrich A. 1945. The use of knowledge in society. *American Economic Review* 35 (4): 519–530.

Hurwicz, Leonid, and Stanley Reiter. 2006. *Designing Economic Mechanisms*. Cambridge University Press.

Kahneman, Daniel, and Amos Tversky. 1979. Prospect theory: An analysis of decision and risk. *Econometrica* 47 (2): 263–291.

Lo, Andrew. 2012. Adaptive markets and the New World Order. *Financial Analysts Journal* 68 (2): 18–29.

Milgrom, Paul, and John Roberts. 1992. *Economics, Organization and Management*. Prentice-Hall.

Myerson, Roger B. 1991. *Game Theory: Analysis of Conflict*. Harvard University Press.

Smith, Adam. 1776. *The Wealth of Nations*. (There are many modern editions.)

Vickrey, William. 1961. Counterspeculation, auctions, and competitive sealed tenders. *Journal of Finance* 16 (1): 8–37.

von Neumann, John, and Oskar Morgenstern. 1944. *Theory of Games and Economic Behavior*. Princeton University Press.

Williamson, Oliver E. 1975. *Markets and Hierarchies*. Free Press.

Recommended Readings

The recommended readings provided for each chapter are works that we feel provide particularly good entry points into each discipline for readers from other disciplines. These readings include classic foundational works, introductory textbooks, and reports of recent research with special relevance for collective intelligence. For this chapter, the references listed above provide a good list of classic foundational works. Below we list some recent works of particular relevance to collective intelligence.

Colin F. Camerer and Ernst Fehr. 2006. When does "economic man" dominate social behavior? *Science* 311 (5757): 47–52.

This paper provides examples of how a variety of experimental results can be explained by extending traditional economic models to include other factors, such as the bounded rationality of human actors and the fact that humans sometimes care about what happens to others, not just to themselves.

Andrew W. Lo. 2013. The origin of bounded rationality and intelligence. *Proceedings of the American Philosophical Society* 157 (3): 269–280.

This provocative article suggests that intelligence can be usefully defined as any behavior that increases the reproductive success of a species. Thus a behavior that appears irrational from the point of view of a single individual may be optimal from the point of view of the whole species. For instance, humans have a well-documented propensity to do "probability matching" in uncertain environments instead of picking the option that would optimize their own individual chances of success. Lo shows, however, that this behavior can be optimal from the point of view of the species if it prevents all individuals in the species from selecting a choice that could, in certain circumstances, lead to the extinction of the whole species.

Scott E. Page. 2007. *The Difference: How the Power of Diversity Creates Better Groups, Firms, Schools, and Societies*. Princeton University Press.

This very readable book summarizes a rich body of research on how differences among the individuals in a group affect the group's ability to solve problems and make accurate predictions. Drawing on economic theorems, computer simulations, and empirical studies of real groups, Page show how, in many situations, groups of people with diverse abilities perform better than groups of the best individual performers.

Justin Wolfers and Eric Zitzewitz. 2004. Prediction markets. *Journal of Economic Perspectives* 18 (2): 107–126.

Prediction markets, in which people buy and sell predictions about uncertain future events, are often a very good way of pooling many people's knowledge to estimate the probabilities of the events. This paper provides a good overview of how these markets work and what is known about them scientifically. A more comprehensive discussion about how groups can estimate probabilities and other quantities can be found in the chapter on Cognitive Psychology in the present volume.

The Wisdom of Crowds vs. the Madness of Mobs

Andrew W. Lo

For at least three reasons, financial markets offer an ideal laboratory in which to study collective intelligence: (1) We have the most and the best data for these markets, virtually all accessible electronically; (2) The stakes are the largest—hundreds of billions of dollars flow through financial markets each day; and (3) Financial markets are known for manias, panics, and crashes, which many perceive as clear failures of collective intelligence.

The success of any market economy depends on collective intelligence across many contexts and over multiple scales. For example, pharmaceutical companies must predict the likely success of various drug-development projects, entrepreneurs must accurately estimate market demand, venture capitalists must accurately assess management skill and product quality, and corporations must correctly anticipate their costs and revenues. In short, most economic activity involves some degree of forecasting, some expectation about the future, and some understanding of how business conditions will evolve, i.e., some form of collective intelligence.

Economics may be unique among the sciences in offering a formal mathematical proof that collective intelligence works, at least in the case of the market. Using minimal assumptions about the behavior of market participants, the famous Arrow-Debreu (1954) model of general equilibrium showed that rational individuals, when left to their own devices, will always "solve" the problem of finding positive prices at which supply equals demand for all goods in the market. This is an extremely difficult problem to solve using other methods, as the former Soviet Union and other command-and-control economies have demonstrated. In fact, the Arrow-Debreu framework apparently "proves too much," since markets seem to function well even under conditions far removed from the framework's basic assumptions of perfect competition and rank-ordered preferences.

Economics is also unique in that it has a compelling hypothesis—known as the Efficient Markets Hypothesis—that prices determined by unconstrained capital markets fully incorporate all available information. Developed independently in 1965 by Paul Samuelson and by Eugene Fama, the Efficient Markets Hypothesis has a counterintuitive Zen-like quality to it: future price changes are unpredictable because if they were predictable, so many greedy investors would attempt to exploit those predictabilities that the predictability would be eliminated almost immediately. Moreover, the information that created the predictability would then be incorporated into market prices, favorable information causing prices to increase and unfavorable information causing prices to decline. Although the formal derivation of the Efficient Markets Hypothesis is highly technical, it provides a concrete and plausible mechanism for how markets are able to incorporate the collective knowledge of their participants effectively.

Fama's (1970) original analysis distinguished three forms of the Efficient Markets Hypothesis: the weak form, the semi-strong form, and the strong form of market efficiency. These forms differ in the amount of information the collective intelligence of the market incorporates. In the weak form, prices reflect all past available public information about an asset. In the semi-strong form, prices reflect all available public information. In the strong form, prices reflect all information about an asset, private and public, including insider information.

For the active investor, the Efficient Markets Hypothesis is something of a wet blanket. In its various forms, it implies that you can't beat the collective intelligence of the market no matter how much you know about an asset, no matter how intelligent you are, and no matter if you possess inside information. Statistically, there will always be someone who has taken advantage of the opportunity to profit before you.

But for the average investor, the Efficient Markets Hypothesis is far from being just a piece of theoretical sleight of hand. This theory was a godsend, paving the way for index funds, passive investing, and the democratization of finance. A consumer or a firm can make informed budgetary or financial decisions about any asset for which there is an efficient market simply by knowing the price of an asset, which, in turn, forms the basis of the modern economy. In the terminology of Malone, Laubacher, and Dellarocas (2010), the collective intelligence of the market is a perfect example of the power of the crowd over the hierarchy and of the value of money over love or glory as a motivator.

The Wisdom of Financial Crowds

Under most conditions, the collective intelligence of the market functions exceedingly well. Markets are robust entities, and the price of an asset carries with it a built-in incentive for people with superior information to profit from their knowledge. As a result, prices quickly converge to incorporate all information on an asset, even in response to the most unexpected occurrences. Maloney and Mulherin's classic 2003 study of the immediate market effects of the *Challenger* disaster demonstrates how powerful the market can be in collecting and applying such information.

On Tuesday, January 28, 1986, just 73 seconds after liftoff at 11:38 a.m. Eastern Standard Time, the Space Shuttle *Challenger* exploded and disintegrated over the Atlantic Ocean. Millions around the world watched this disaster unfold in real time on television. Because the mission included a schoolteacher—Christa McAuliffe, the Shuttle's first civilian passenger—millions of American schoolchildren were watching. The vast majority of Americans learned about the incident within an hour despite the absence of the Internet, Twitter, and other forms of instant communication that now connect us.

Six days after the disaster, President Ronald Reagan signed Executive Order 12546 establishing the Rogers Commission to investigate. On June 6, 1986, after conducting scores of interviews with all the major stakeholders involved in the Space Shuttle program, analyzing all the telemetry data from *Challenger* during its short flight, sifting through the physical wreckage, and holding several public hearings, the Rogers Commission concluded that the explosion had been caused by the failure of "O-rings" on the right-side solid-fuel booster rocket.[1]

Four major contractors were involved in the Space Shuttle program: Lockheed, Martin Marietta, Morton Thiokol, and Rockwell International. The release of the Rogers Commission report was bad news for Morton Thiokol, the private-sector contractor that had been hired to build and operate the booster rockets. However, it must have been a welcome relief for the other three companies, who were now in the clear after four months of rumors, finger pointing, and government investigation.

The Rogers Commission took a little more than four months to determine the cause of the *Challenger* disaster. How long did it take for the collective intelligence of the market to process the *Challenger* explosion and incorporate it into the stock prices of the four vendors? Maloney and Mulherin concluded that it took less than thirteen minutes.

The price of Morton Thiokol stock began to decrease almost immedi-ately after the accident. By 11:52 a.m., only thirteen minutes after the explosion, the New York Stock Exchange had to halt trading in Morton Thiokol—and only in Morton Thiokol—because the flow of orders had overwhelmed the exchange's systems. By the time trading in Morton Thiokol stock resumed, later that afternoon, its price had decreased by 6 percent; by the end of the day, it was down almost 12 percent—a deep statistical outlier relative to its past performance (see figure 2.1). The trading volume of Morton Thiokol shares on January 28, 1986 was seventeen times the previous three-month average. Although the stock prices of Lockheed, Martin Marietta, and Rockwell International also experienced declines on that day, their decreases in price and in overall volume traded were much smaller and were within statistical norms.[2]

A cynical reader might suspect the worst: perhaps engineers at Morton Thiokol or at NASA recognized what had happened and began dumping their stocks immediately after the accident. But Maloney and Mulherin were unable to find evidence for gross insider trading on January 28, 1986. Even more startling was the fact that the lasting decline in the market capitalization of Morton Thiokol on that day—about $200 million—was almost exactly equal to the damages,

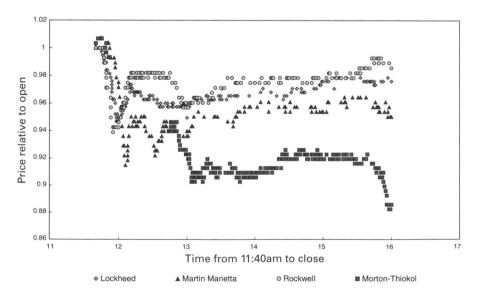

Figure 2.1
Intradaily prices of Lockheed, Martin Marietta, Rockwell International, and Morton Thiokol immediately after the *Challenger* explosion on January 28, 1986. Source: Maloney and Mulherin 2003.

settlements, and lost future cash flows that Morton Thiokol later incurred. The collective intelligence of the stock market was able to do within a few hours what took the Rogers Commission four months.

The Madness of Mobs

As powerful as the collective intelligence of the market is, it is far from infallible. On May 6, 2010, beginning at about 2:32 p.m. Eastern Daylight Time, the Dow Jones Industrial Average took a nosedive that turned into its biggest one-day point decline on an intraday basis in its 126-year history: 998.5 points. This "Flash Crash" lasted a little over half an hour, during which the stock prices of some of the world's largest companies traded at absurd prices. Accenture fell to a penny a share (see figure 2.2), while Apple briefly shot up to $100,000 per share.

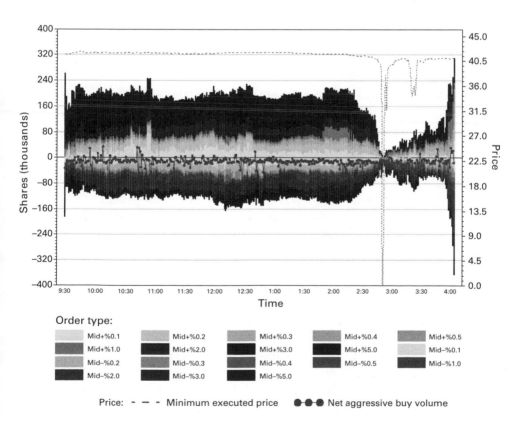

Figure 2.2
Price and order-book depth and net aggressive buy volume for Accenture plc (ACN) on May 6, 2010. Source: CFTC/SEC 2010, table 4.A. Reproduced without color.

At the end of this half-hour, prices returned to their previous levels, as if nothing had happened and it was just a prank.

How did the wisdom of crowds turn so quickly into the madness of mobs? The short answer is that even now, five years after this fiasco, nobody knows for sure. After five months of intensive investigation, the Commodity Futures Trading Commission (CFTC) and the Securities and Exchange Commission (SEC) issued a joint report concluding that these events had occurred not because of any single failure, but rather as a result of seemingly unrelated activities across different parts of the financial system—an instance of collective ineptitude.[3] Their joint report ascribed the failure to a deadly combination of four seemingly innocuous things: a single mutual fund's automated execution algorithm on autopilot, a game of "hot potato" among high-frequency traders, cross-market arbitrage trading, and the use of "stub quotes" (automated placeholder price quotes so far from the range of normal trading that no one expected them to be executed, but a number of them did get executed inadvertently on May 6, 2010). Apparently, all these things conspired to create a brief but perfect financial storm of extreme price volatility.

Although such a narrative seems compelling, several financial institutions have presented empirical evidence that contradicts certain aspects of the joint report, including data that are inconsistent with the claim that the initial cause of the Flash Crash was a rapid automated sale of 75,000 S&P 500 E-mini June 2010 stock-index futures contracts, worth approximately $4.1 billion, on the Chicago Mercantile Exchange (CME) by a single mutual-fund complex as a hedge for its equity position.[4] In fact, as this chapter undergoes its final editing a London-based day trader has been arrested and is being extradited to the U.S. to face criminal and civil charges for allegedly engaging in fraudulent trading activity that led to Flash Crash (Anderson 2015).

What is not in dispute is that sharp price declines over a 20-minute period in some of the largest and most heavily traded stocks cascaded into a systemic shock for the entire U.S. financial market. Market makers found that the liquidity of some stocks was utterly exhausted. And after all human traders withdrew their quotes from the market-place, what remained were stub-quote placeholders, which led to the temporary penny-stock valuation of Accenture and a valuation of $100,000 a share for Apple when some of these quotes were executed. As a result of the Flash Crash, 5.5 million shares were traded at prices that were more than 60 percent different from their pre-crash values.

(These trades were later canceled by exchange authorities, but trades that were completed at 59 percent away from their pre-crash values were not canceled.[6])

Malone, Laubacher, and Dellarocas (2010) have emphasized the importance of the correct incentives to produce collective intelligence. In economics, the maximization of profit is held to be a sufficient incentive to produce the correct market price. After all, any opportunity to make a profit will quickly be arbitraged away by market participants. However, during the Flash Crash the incentive to maximize profit was overwhelmed and even subverted by market dynamics. Just as with paid crowdsourcing, the profit incentive sometimes isn't sufficient to produce the desired results (Bigham, Bernstein, and Adar, this volume), and it wasn't enough to maintain the collective intelligence mechanism of the market during this crisis.

The Flash Crash allows us to trace this failure in collective intelligence as it happened. Information about the price of an asset in a market is incorporated through the buying and selling of the asset, so any discrepancy in price is exploited to make a profit. In the Flash Crash, however, cross-market arbitrage transmitted the rapid decline in E-mini price to other markets, since the E-mini was traded only on the CME. These cross-market arbitrageurs traded similar assets on other exchanges, such as SPYs, or created "synthetic" versions with baskets of individual securities. This was not aberrant behavior on the part of the cross-market arbitrageurs. Had the E-mini plunged for legitimate reasons, these arbitrage traders would have been the means through which information was transmitted to other financial markets. However, the growing interconnectedness of financial markets has also created a new form of financial accident: a systemic event in which the "system" extends beyond any single market and affects a great number of innocent bystanders.

The Financial Crisis of 2007–2009

The largest systemic failure of the collective intelligence of the market of the last eighty years is, of course, the housing bubble and its collapse, which precipitated the financial crisis of 2007–2009. The collective intelligence of the market left very little room for dissenting opinions about the housing market during the bubble, even though those opinions were a well-disseminated part of the public record. The lack of diversity in market opinion led to prices that reflected optimism about the housing market far more than information about obscure potential risks in the tranches of mortgage-based securities. At every step of the mortgage financing process, the incentives for participants were skewed in favor of a belief in the appreciation of housing value, despite growing evidence to the contrary. Every type of stakeholder in the housing boom benefited from the real-estate market's upward trend: homeowners, commercial banks, government-sponsored enterprises, credit-rating agencies, insurance companies, investment banks, hedge funds, pension funds, mutual funds, mortgage lenders, mortgage brokers, mortgage servicers, mortgage trustees, politicians, regulators, and even economists. "It is difficult," the novelist Upton Sinclair once remarked, "to get a man to understand something, when his salary depends upon his not understanding it." But consider how much more difficult it is to convince someone that doom is around the corner when the money is still coming in.

The price discovery mechanism that leads to the collective intelligence of the market depends on investors taking advantage of profit opportunities to incorporate new information into prices. Yet during the housing bubble this process failed to incorporate information in a timely enough manner to prevent severe economic dislocation. It was very difficult to "bet" against the trend. At various times, attempts to "short" the housing bubble were met with open disbelief. The financial journalist Gregory Zuckerman (2009) recounts how sellers of credit default swaps on subprime-mortgage-based securities tried to warn John Paulson (one of the very few people who managed to profit from the collapse of the housing bubble) not to purchase them, since he would lose money on the trade. Paulson personally netted $4 billion in 2007 as a result of that bet.

The collective intelligence of the market produces information only about the price of an asset. Every scrap of greater context is assumed to have been arbitraged and incorporated into that price. However, during the housing bubble, this reliance on price without greater context fostered an unawareness of the hidden interdependencies within the market—for example, the degree of securitization of residential mortgages, which surprised the vast majority of economists upon its "discovery," and about which the average borrower knew absolutely nothing. Under normal circumstances, this simplification of a complex financial process into a single number is an important strength of the collective intelligence of the market. However, during the collapse of the housing bubble, when many borrowers were caught unaware by the new terms and dispositions of their mortgages, that simplification became a weakness.

Moore's Law vs. Murphy's Law

Despite its failures, both small and large, the collective intelligence of financial markets is still an overwhelming success story. From 1929 to 2009, the total market capitalization of the U.S. stock market has doubled every ten years. The total trading volume of stocks in the Dow Jones Industrial Average doubled every 7.5 years during this period, but since 2003 the pace has accelerated: now the doubling occurs every 2.9 years, growing almost as fast as the semiconductor industry. Like every other industry that has reduced its costs via automation, the financial industry has been transformed by information technology. There is no doubt that algorithmic trading has become a permanent and important part of the financial landscape, yielding tremendous cost savings, operating efficiency, and scalability to every financial market it touches.

However, stock exchanges, telecommunications networks, and automated trading algorithms are just examples of tools designed by humans and used by humans to pursue human goals, the modern equivalents of the flint blade, the tribal unit, and the printing press. And the carelessly calibrated stub quotes at the epicenter of the Flash Crash were not freaks of nature; they were programmed by human actors. We are accustomed to technology that allows us to work at scales outside the range of normal human perception in other fields, from microsurgical techniques to the arrays of astronomical instruments that allow us to confirm the predictions of Einstein's theory of general relativity. The acceleration of market dynamics is just one more example of the natural tendency of humankind to extend its reach into previously uncharted territory.

Any technology that supersedes human abilities has the potential to create unintended consequences. Financial technology provides enormous economies of scale and scope in managing large portfolios, but now trading errors can accumulate losses at the speed of light before they can be discovered and corrected by humans. Obviously, with the increase in the speed of financial markets, the frequency of technological malfunctions, price volatility spikes, and spectacular frauds and failures of intermediaries has increased noticeably. The benefits of Moore's Law can quickly be offset by the costs of Murphy's Law.

But financial crises also say something deeper about two important issues in collective intelligence: diversity and hierarchy. Woolley et al. (2010) have demonstrated that in small groups the level of collective intelligence increases with the social sensitivity of a group's members, with a relatively equal distribution of conversational turn taking, and with the proportion of women in a group. This suggests that collective intelligence may function best in relatively diverse and egalitarian settings in which many different opinions are represented and heard, although the specific type of diversity is still an open question (Woolley, Aggarwal, and Malone, this volume).

There has been an unspoken tradeoff between speed and diversity in automated financial markets. The logic is easy to understand. Because prices are volatile, market participants want their orders executed as quickly as possible. They want to profit from their insight into the market, whether it is real or illusory, before the price changes. This requires financial intermediaries to maintain a continuous presence in a market—a role once held by human market makers and specialists. Automation of the trading process, including computerized algorithmic trading, has drastically reduced the costs to the intermediaries of maintaining a continuous market presence. In fact, the intermediaries with the most efficient trading technology and the lowest regulatory burden realized the largest cost savings. As a result, the supply of immediacy in financial markets has skyrocketed.

However, the net benefits of immediacy have accrued disproportionally to those who can better absorb the costs of participating in automated markets. This has frustrated and disenfranchised a large population of smaller and less technologically advanced market participants. The consequences of this lack of diversity are apparent in the "hot potato" phase of the Flash Crash. According to the CFTC/SEC joint report, after the imbalance in the S&P 500 E-mini futures market (allegedly caused by the rapid sale of 75,000 E-mini contracts, though this is now in dispute), approximately 50,000 contracts found net buyers. But sixteen high-frequency traders passed 27,000 E-mini contracts back and forth, at one point accounting for 49 percent of the CME's trading volume, finding buyers for only 200 of them. This algorithmic groupthink, instead of providing liquidity for the market, apparently competed for liquidity to the exclusion of other market participants, driving down the price of the E-mini contracts even further during the Flash Crash. To make an analogy to the collective intelligence of human teams, it was as if a small clique with a bad idea had hijacked the price discovery process for its own purposes. And if the current charges against the London-based day trader prove true, it means that one hijacker was able crash the largest stock market in the world using off-the-shelf software from the comfort of his own home.

The Flash Crash also illustrates the importance of organization in the proper functioning of collective intelligence. The market is a crowd-sourced solution to the specialized and mathematically challenging problem of determining asset prices. But through the dynamics of the Flash Crash, we can see that the price discovery process was not a collaboration of equals in a crowd. Small groups of market participants commandeered this process in previously unforeseen ways. Just as the collective intelligence of the market successfully identified Morton Thiokol as the biggest economic casualty of the *Challenger* disaster in a matter of minutes, failures of collective intelligence reveal emergent hierarchies and interdependencies in the market.

An Evolutionary Perspective

Technology is reshaping the collective intelligence of the market at an unprecedented pace. Although information technology has advanced tremendously in the last few decades, human cognitive abilities have been largely unchanged over the last several millennia. Humans have been pushed to the periphery of a much faster, larger, and more complex trading environment. Indeed, the enhanced efficiency, precision, and scalability of algorithms may diminish the effectiveness of those risk controls and systems safeguards that rely on experienced human judgment and are applied at human speeds. Technology will continue to change the market environment, and market participants will either adapt or go extinct. As we'll see, this is an important feature of the collective intelligence of the market.

Markets are a form of "crowd" collective intelligence. Extrapolating from Woolley's results, an efficient crowd requires a diversity of opinion to function more effectively than the sum of its parts. Similarly, a market requires diversity of opinion to function at all—it requires both buyers and sellers. To ensure that this happens, some financial markets require a market maker to buy and sell a given security, maintaining both sides to provide liquidity to the market, as when a crowd-sourced problem-solving effort deliberately includes a contrarian or a devil's advocate in a group. Market making is a risky activity because of price fluctuations and adverse selection—prices may suddenly move against market makers and force them to unwind their positions at a loss. Similarly, being the voice of dissent in a collective intelligence might be uncomfortable for that group member, even if it helps the group as a whole to find a solution.

But this leads to a greater point: Market dynamics are, in fact, population dynamics. Failures in the collective intelligence of the market can be traced to particular pathologies in the dynamics of market participants. For example, a "crowded trade" is a trade that has a large number of investors (usually short-term investors) crowding into an asset on the belief that it will appreciate quickly. When it doesn't, everyone rushes to the exit doors at the same time, leading to even greater losses. However, for reasons that are still debated, the summer of 2007 ushered in a new financial order in which the crowded-trade phenomenon applied to entire classes of portfolio strategies—most notably in the "Quant Meltdown"—not just to a collection of overly popular securities.[7] A monoculture of financial behavior is unhealthy in a market, whether among high-frequency trading programs, hedge funds' statistical arbitrage strategies, or first-time home buyers, just as a monoculture of opinion is unlikely to exhibit collective intelligence.

Understanding the behavior of market participants is essential to understanding the collective intelligence of the market. Even in the abstract microsecond level of algorithmic trading, we can still classify market behaviors in terms of human motivation. In the Flash Crash, stub quotes were the reason prices changed so dramatically as liquidity was withdrawn from the market, and these price changes led to further trades executed at even more extreme prices as investors reacted to these shocks. However, humans ultimately chose the parameters under which these orders were executed. It was indistinguishable from a panic sale, and much of the market plunge was a reaction to that event.

At the same time, the increased speed of order initiation, communication, and execution has become an opportunity for the fastest market participants to profit. Given these profit opportunities, some participants may choose to engage in a "race to the bottom," forgoing certain risk controls that slow down order entry and execution. (This may explain what happened among the high-frequency traders during the game of "hot potato" in the Flash Crash.) The result can be a vicious cycle of misaligned incentives as greater profits accrue to the fastest market participants with the least comprehensive safeguards. In human terms, this would be the rough equivalent of a soldier's taking a swig of whiskey before entering battle—an understandable reaction to the short-term terror of armed conflict but not necessarily conducive to long-term survival. However, neurologists already know what happens financially to humans who have lost their sense of fear but otherwise appear "rational": they take risks incommensurate with potential gains, and quickly go bankrupt.[8]

Investors are always poised on a delicate balance between greed and fear. Economists call this balance "rationality," but neuroscientists have known for decades that economic rationality emerges from the interaction of brain components usually identified with the processing of emotion. This is, of course, a form of collective intelligence at a neurophysiological level. Traders who exhibit too little or too much emotional response tend to reap smaller profits than those with mid-range sensitivity.[9] Market dynamics reflect the neuropsychology of groups of market participants. The infamous film character Gordon Gekko said "Greed is good,"; in fact, neuroscientists have shown that monetary gain stimulates the same reward circuitry in the nucleus accumbens as cocaine. Apparently, greed feels good, too.

And the growth of algorithmic trading has not prevented traders from feeling gains and losses very keenly indeed. During the "Quant Meltdown" of August 2007, I was contacted by several former students who were experiencing "shell-shock" as a result of the mounting losses incurred by their firms' statistical arbitrage strategies. Many firms, afraid of further losses, changed their positions—and lost twice when the market suddenly reversed itself at the conclusion of the Quant Meltdown. The basic human emotions of fear and greed can still have profound effects on even the most technologically advanced forms of automated trading. The collective intelligence of the market is still prone to wild swings of emotion—to what John Maynard Keynes called "animal spirits."

Adaptive Markets and Collective Intelligence

With these facts in mind, a counter-hypothesis to the Efficient Markets Hypothesis can be formulated to explain the collective intelligence of the market—a counter-hypothesis that takes the natural variation of market behavior into account. The classical form of the Efficient Markets Hypothesis tells us that prices naturally reflect all available information due to the relentless pressure of market participants trying to make a profit on even the slightest informational advantage. However, assuming that populations of market participants always behave to maximize their profit is simplistic. In practice, market participants show a complicated and dynamic medley of several decision-making heuristics, ranging from primitive hardwired responses such as the fight-or-flight reflex to the more refined mental processes that we traditionally associate with rationality and profit maximization. Financial markets and their denizens are not always rational, nor are they always emotional.

Instead, under the Adaptive Markets Hypothesis (Lo 2004, 2005, 2012), they are constantly adapting to new sets of circumstances in an evolutionary process.

Like a good theory, the Adaptive Markets Hypothesis can explain phenomena that its predecessor cannot, but also reduces to the Efficient Markets Hypothesis when market environments are ideal or "frictionless." However, market environments are rarely ideal. The Adaptive Markets Hypothesis is based on the principles of evolutionary biology: the process of natural selection rewards successful participants in a market environment and punishes unsuccessful ones, leading to changes in market behavior and populations over time. If market environments are relatively stable, the process of natural selection will eventually allow market participants to successfully adapt to this period of market stability, after which the market will appear to be quite efficient. If the environment is unstable, however, market dynamics will be considerably less predictable as some "species" lose their competitive edge to others. In this case, the market may appear to be quite inefficient. Deborah Gordon's example of colonies of harvester ants that change their behavior as their environment experiences drought (see the chapter on biology in this volume) is a very close biological analogy to this market process. The Adaptive Markets Hypothesis explains the dynamic behavior of these financial species, from the macroeconomic scale of major mutual funds to the microsecond scale of high-speed trading. Indeed, the evolutionary theory of "punctuated equilibrium," in which infrequent but large environmental shocks can cause massive extinctions followed by a burst of new species, is directly applicable to the pattern of extinctions and innovations in the world of hedge funds and proprietary trading.

The Adaptive Markets Hypothesis also tells us that the collective intelligence of the market does not exist in a vacuum; it is non-stationary, highly institution dependent, and highly path dependent. Extended periods of prosperity have an anesthetic effect on the human brain, lulling investors, business leaders, and regulators into a state of complacency that can lead them to take risks they might otherwise avoid. The wake-up call, when it comes, is just as extreme. Once the losses start mounting, the human fear circuitry kicks in and panic ensues, a flight-to-safety that often ends with a market crash. As Bernstein, Klein, and Malone (2012) note, the "chaotic dynamics" of bubbles and crashes have a natural and all-too-familiar progression firmly rooted in the functioning of the human brain.

In light of these human frailties, it is no surprise that since 1974 there have been at least seventeen banking-related national crises around the world in which the collective intelligence of the market clearly failed. The majority of these events were preceded by periods of rising real-estate and stock prices, large inflows of capital, and financial deregulation, producing an unwarranted sense of financial euphoria. However, the Adaptive Markets Hypothesis also suggests that collective intelligence will emerge most effectively when its participants are emotionally well calibrated—that is, neither emotionally underinvested nor overinvested in any given project.

Conclusion

Why has the collective intelligence of the market been so successful, despite its occasional failure? The Adaptive Markets Hypothesis tells us that profit taking alone is not enough to explain market success. The adaptive behavior of the market uses the power of positive and negative selection to drive the evolutionary process. The market rewards participants who help further the price discovery process (who then are copied by others in the market), but it also punishes participants who are mistaken in their price conjectures. These rewards and punishments are, by definition, meted out in exact proportion to their value to the market. The collective intelligence of the market is efficient in price discovery precisely because it selects for an efficient outcome. Although this set of incentives and disincentives can be mimicked to some extent outside of a market context by mechanisms such as up-and-down voting in reputational social media, the elegance of the market mechanism is the simplicity and effectiveness of a process that is instantaneous, non-hierarchical, and almost entirely self-organized and autonomous.

Imagine this mechanism being used in other forms of collective intelligence—for example, imagine a version of Wikipedia in which changes were also rewarded or punished automatically according to their value. In fact, the process of scientific peer review approximates this practice, though much more slowly—but if there has been one form of collective intelligence in the modern era more successful than the market, it has been the process of scientific discovery. Accordingly, the chapters in this volume, and the impact that their contents may or may not have in the coming years, will be a live experiment of collective intelligence at work.

Acknowledgments

I thank Michael Bernstein, Jayna Cummings, Tom Malone, and anonymous reviewers for helpful comments and discussion. Research support from the MIT Laboratory for Financial Engineering is gratefully acknowledged.

Notes

1. See the Report of the Presidential Commission on the Space Shuttle Challenger Accident (1986).

2. See Maloney and Mulherin 2003.

3. See CFTC/SEC 2010.

4. See CME 2010 and Nanex 2014 for detailed rebuttals to the CFTC/SEC 2010 joint report on the Flash Crash. For the recent arrest of the London day trader, see Viswanatha, Hope, and Strasburg 2015.

5. See CFTC/SEC 2010, p. 17.

6. See ibid., pp. 63–65.

7. See Khandani and Lo 2007, 2011.

8. See Bechara et al. 1994.

9. See Lo and Repin 2002 and Lo, Repin, and Steenbarger 2005.

References

Anderson, Jenny. 2015. Trader charged in 'Flash Crash' case to fight extradition. *New York Times*, April 22.

Arrow, Kenneth. J., and Gérard Debreu. 1954. Existence of an equilibrium for a competitive economy. *Econometrica* 22: 265–290.

Bechara, Antoine, Antonio R. Damasio, Hanna Damasio, and Steven W. Anderson. 1994. Insensitivity to future consequences following damage to human prefrontal cortex. *Cognition* 50 (1–3): 7–15.

Bernstein, Abraham, Mark Klein, and Thomas W. Malone. 2012. Programming the global brain. *Communications of the ACM* 55 (5): 41–43.

CFTC/SEC (Commodity Futures Trading Commission/Securities and Exchange Commission). 2010. Preliminary Findings Regarding the Market Events of May 6, 2010. Report of the Staffs of the CFTC and SEC to the Joint Advisory Committee on Emerging Regulatory Issues. http://www.sec.gov/sec-cftc-prelimreport.pdf.

CME (Chicago Mercantile Exchange). 2010. CME Group Statement on the Joint CFTC/SEC Report Regarding the Events of May 6.

Fama, Eugene F. 1970. Efficient capital markets: A review of theory and empirical work. *Journal of Finance* 25: 383–417.

Khandani, Amir E., and Andrew W. Lo. 2007. What happened to the quants in August 2007? *Journal of Investment Management* 5 (4): 5–54.

Khandani, Amir E., and Andrew W. Lo. 2011. What happened to the quants in August 2007? Evidence from factors and transactions data. *Journal of Financial Markets* 14 (1): 1–46.

Lo, Andrew W. 2004. The adaptive markets hypothesis: Market efficiency from an evolutionary perspective. *Journal of Portfolio Management* 30: 15–29.

Lo, Andrew W. 2005. Reconciling efficient markets with behavioral finance: The adaptive markets hypothesis. *Journal of Investment Consulting* 7: 21–44.

Lo, Andrew W. 2012. Adaptive markets and the new world order. *Financial Analysts Journal* 68: 18–29.

Lo, Andrew W., and Dmitry V. Repin. 2002. The psychophysiology of real-time financial risk processing. *Journal of Cognitive Neuroscience* 14: 323–339.4

Lo, Andrew W., Dmitry V. Repin, and Brett N. Steenbarger. 2005. Fear and greed in financial markets: An online clinical study. *American Economic Review* 95: 352–359.

Malone, T. W., R. Laubacher, and C. Dellarocas. 2010. The collective intelligence genome. *MIT Sloan Management Review* 51 (3): 21–31.

Maloney, Michael T., and J. Harold Mulherin. 2003. The complexity of price discovery in an efficient market: The stock market reaction to the *Challenger* crash. *Journal of Corporate Finance* 9: 453–479.

Nanex, LLC. 2014. Reexamining HFT's Role in the Flash Crash.

Report of the Presidential Commission on the Space Shuttle Challenger Accident. 1986. U.S. Government Printing Office.

Viswananatha, Aruna, Bradley Hope, and Jenny Strasburg. 2015. Flash Crash charges filed. *Wall Street Journal*, April 21.

Woolley, Anita Williams, Christopher F. Chabris, Alex Pentland, Nada Hashmi, and Thomas W. Malone. 2010. Evidence for a collective intelligence factor in the performance of human groups. *Science* 330: 686–688.

Zuckerman, Gregory. 2009. *The Greatest Trade Ever: The Behind-the-Scenes Story of How John Paulson Defied Wall Street and Made Financial History*. Crown.

Biology

Editors' Introduction

Biologists have studied collective intelligence in organisms for at least as long as social scientists have been studying it in humans. In some sense, as Aristotle pointed out, evolution itself can appear intelligent when we admire how well natural systems function in relation to one another. Ants, fish, and even slime molds all demonstrate astonishing adaptations to survive and thrive. Whereas in human organizations rules are often communicated from the top down, natural systems tend to follow distributed rules that build up globally coherent behavior through local interactions. Operating without central control or leadership, many organisms still take collectively intelligent actions, effectively exploring their environments and maximizing their effectiveness in them. Organisms' effectiveness without central control suggests similarities to the financial markets that Andrew Lo discussed in the preceding chapter.

As Deborah Gordon describes in the chapter that follows, biologists investigate collective behavior through a combination of lab science, field observation, modeling, and simulation. Lab and field investigations use observations to produce new models of the organism. A scientist then often tests the predictions of these models with further observations. Biological studies of collective behavior have the advantage that some natural systems can be observed in their entirety. It is difficult to observe an entire human city, but somewhat easier to track an ant colony's evolution or a slime mold's growth.

Evolution has shaped natural systems to function in particular environments. This is true of collectives that involve independent individuals, such as flocks of birds and colonies of ants, and of smaller collectives, such as bacteria. These organisms provide many diverse examples of collective intelligence—indeed, in some ways the examples they provide may be more diverse than those discussed in any of this volume's other chapters.

Gordon focuses on ecology, which studies the interactions between organisms and their environments. How do ants decide how many members of the colony to send out in search of sustenance? How do fish and birds develop and coordinate collective behavior such as flocking? Organizations face many similar questions; thus, the chapter on organizational behavior serves as a counterpoint to the perspectives here.

In addition to the work summarized in this chapter, the interdisciplinary field of neuroscience—which overlaps heavily with biology—is also very relevant for collective intelligence. For instance, questions about how different neurons and different parts of the brain work together to produce intelligent behavior are central to neuroscience (see, e.g., Seung, 2012), and their answers may have implications for designing or understanding other kinds of collective intelligence. Some work, for example, shows how the mental processing people do to understand other people's behavior occurs in a specific region of the brain called the temporo-parietal junction (Saxe and Kanwisher 2003).

Gordon synthesizes years of research into the collective efforts of organisms such as ants. It offers colorful analogies between the collective behavior of ants and the Internet's TCP backbone algorithm, human brain activity, and streams of water. In fact, it suggests the possibility of articulating generalizable motifs, or patterns, in system configurations. Malone, Laubacher, and Dellarocas (2010) suggest a similar approach to identifying common patterns in new Internet-enabled forms of human collective intelligence. We thus believe that cross-disciplinary connections such as those discussed in Gordon's chapter and in other parts of biology can be more than just colorful analogies. We believe that the generalizable patterns they embody can help us to apply insights from biology to human systems, and vice versa.

References

Malone, T. W., R. Laubacher, and C. Dellarocas. 2010. The collective intelligence genome. *MIT Sloan Management Review* 51 (3): 21–31.

Saxe, R., and N. Kanwisher. 2003. People thinking about thinking people: The role of the temporo-parietal junction in "theory of mind." *NeuroImage* 19: 1835–1842.

Seung, S. 2012. *Connectome: How the Brain's Wiring Makes Us Who We Are*. Houghton Mifflin Harcourt.

Recommended Readings

Deborah M. Gordon. 2010. *Ant Encounters: Interaction Networks and Colony Behavior*. Princeton University Press.

Deborah Gordon's book provides an accessible, deep treatment of many of the phenomena she introduces in her chapter in the present volume. In *Ant Encounters*, she asks how ant colonies can get things done with no individual in charge. Her observations focus on the emergent results of local interactions between individual ants in the colony.

Andrew Berdahl, Colin J. Torney, Christos Ioannou, Jolon J. Faria, and Iain D. Couzin. 2013. Emergent sensing of complex environments by mobile animal groups. *Science* 339: 574–576.

Schooling fish typically travel in packs. But need they do so? Berdahl et al. discuss how small groups of golden shiners reacted poorly to a rapidly changing environment, but, as the size of the groups increased past a certain number, the group suddenly became collectively much more effective. They show that, as in many of the biological systems mentioned in the chapter below, a group of similarly "unintelligent" organisms can regulate their behavior to adjust to changing conditions even when individuals on their own may have little ability to act effectively.

Balaji Prabhakar, Katherine N. Dektar, and Deborah M. Gordon. 2012. The regulation of ant colony foraging activity without spatial information. *PLoS Computational Biology* 8.8: e1002670.

This paper posits striking connections between biology and computer science and between engineered systems and evolved ones. How do ant colonies regulate how many ants to send outside the nest to forage for water? The ants must expend water in order to find more of it, at risk of finding nothing. The researchers model how a colony can assess food availability. The more frequently ants return with food, the more ants get sent out. This algorithm bears a striking similarity to the Internet's TCP congestion control algorithm, which uses the frequency of a receiving computer's acknowledgments to decide how quickly to send the next packet of data.

Seung, Sebastian. 2012. *Connectome: How the Brain's Wiring Makes Us Who We Are*. Houghton Mifflin Harcourt.

This book, which is aimed for popular audience, provides a good introductory overview of what we know about neuroanatomy and how it relates to the behavior of brains. The book also makes the case for a research approach to understanding how brains work by creating extremely detailed maps of the connections among neurons ("connectomes").

Atsushi Tero, Seiji Takagi, Tetsu Saigusa, Kentaro Ito, Dan P. Bebber, Mark D. Fricker, Kenji Yumiki, Ryo Kobayashi, and Toshiyuki Nakagaki. 2010. Rules for biologically inspired adaptive network design. *Science* 327 (5964): 439–442.

Human societies and biological systems share a reliance on transportation networks. Transportation networks are, of course, extremely difficult to plan effectively. Even simple organisms such as slime molds, however, seem to find near-optimal solutions to this difficult problem. This paper included a memorable demonstration: placing slime-mold food where major cities were shown on a map of Japan prompted a slime mold to grow a network that looked extremely similar to the Tokyo rail system. This result raises an interesting question: Can the behavior of simple organisms help us to plan our own collective behavior?

David J. T. Sumpter. 2010. *Collective Animal Behavior*. Princeton University Press.

This book provides a broad overview of collective behaviors and the mechanisms behind them. The intended audience includes both behavioral ecologists and the broader scientific community. Some of the evidence is from observations, some from experiments, and some from modeling.

Collective Behavior in Animals: An Ecological Perspective

Deborah M. Gordon

Systems without central control are ubiquitous in nature. Cells act collectively—for example, networks of neurons produce thoughts and sensations, and T cells and B cells together mobilize immune response to pathogens. Many animals, too, act collectively—for example, flocks of birds turn in the sky, and schools of fish swerve to avoid predators. Social insects live in colonies, and groups work together to collect food, build nests, and care for the young.

Nature offers a huge variety of examples of collective behavior in animals. (For a review, see Sumpter 2010.) Because hierarchy is familiar in human organizations, it may at first seem surprising that animal groups can function without any central control. Much of the study of collective behavior in animals has been devoted to establishing that there really is no leader, and to considering how this could possibly be true. For example, in early experiments on the formation of a V by a flock of geese, the investigator shot the goose at the front of the V. Another goose immediately took its place, demonstrating that the V shape was not the result of leadership on the part of a particular animal but instead arose from interactions among all the geese, regardless of each goose's personal qualities.

Modeling Animal Collective Behavior

Most of the work done so far on collective behavior in animals seeks to describe how groups of a particular species accomplish a certain task: how fish schools turn, or how ants recruit others to food. The answer to this question is an algorithm that explains, in general, how local interactions produce a particular outcome.

Recent work has made significant progress in identifying the algorithms used in animal collective behavior (Sumpter 2010). Over the past twenty years, across all the fields of biology, attention has turned to deciphering how local interactions create global outcomes. The goal of the first models of collective behavior was to demonstrate that such behavior, called "emergence" or "self-organization," can exist, by showing how it works in a particular case (see, e.g., Deneubourg et al. 1986). A growing field of research continues this effort, seeking to describe the local interactions that give rise to some collective outcome (see, e.g., Couzin et al. 2005). Patterns of animal movement are the best-studied form of collective behavior in animals. Models of the movement of fish schools explain how local interactions among fish allow the school to turn, without any decision from any individual about where to go (Handegard et al. 2012), and studies of the movement of groups of locusts explain how interactions between individuals push the group from the stationary to the migratory phase (Bazazi et al. 2012).

So far the study of collective behavior in animals has been devoted to discovering the algorithms that allow local interactions to produce collective outcomes. Enough progress has been made that it now may be possible to sort types of collective behavior according to the algorithms that generate the behavior.

One set of models, mainly applied to social insects, describe collective behavior as "self-organization" (Nicolis and Prigogine 1977). These models grew from a search for analogies in animals to transitions shown in thermodynamics by Prigogine—examples of collective behavior in which accumulated interactions lead to an irreversible outcome. All that is needed is some local interaction in which one individual influences the behavior of another, such as following, repelling, or imitating. Pour some water on the counter and small streams will lead from the main source and find a way to drip off the edge. The direction the streams will take is a result of local interactions among molecules: when enough molecules go in the same direction to overcome the surface tension of the water, a stream will form. In the same way, ants that use trail pheromone will form trails to resources. From a central nest surrounded by scattered resources, the direction that trails will take depends on local interactions among ants. In order to get started, there must be some random movement by searching ants. If some ants put down a signal, such as a chemical trail, when they find food, then eventually trails will form in the directions that happened to attract enough ants to pull in other ants that then get to the food and reinforce the trail in time to attract others.

Another class of algorithms comes from networks in which the rate of interaction, rather than any particular message conveyed in an interaction, is the cue that provides information in a network. For example, harvester ants use simple positive feedback, based on the rate of interaction between outgoing foragers and foragers returning with food, to regulate foraging activity (Gordon et al. 2011, Pinter-Wollman et al. 2013).

Does all collective behavior in animals qualify as collective intelligence? Sometimes we call collective behavior "intelligent" when it seems to serve a purpose. As one watches ants, it's clear that ants can be quite limited, yet colonies manage to adjust to appropriately to subtle changes in conditions, and ants have been extremely successful worldwide for many millions of years. Sometimes we call collective behavior "intelligent" when it seems to be an example of behavior that we already think is intelligent, such as memory or learning. This follows Turing's lead; we can agree that anything is a brain if it performs particular functions. But to avoid tautology, it is important to keep in mind that the resemblance to intelligence is established by definition. For example, we could ask whether ant colonies learn or remember, and the answer will depend mostly on whether we define learning or remembering to include some things that ant colonies do. In what follows, I will use "collective behavior" interchangeably with "collective intelligence."

We do not yet have a general theory of collective behavior in animals. The growing body of work that shows how local interactions produce collective behavior feeds the hope that there may be general laws that could explain how groups work together. Similar algorithms have been used to describe collective behavior in disparate systems. Systems biologists have noticed that some network configurations, or motifs, occur more often than others, and that these may be due to basic principles that make certain configurations more efficient (Doyle and Csete 2011; Alon 2007). Agent-based models draw on the results in statistical mechanics that predict the behavior of a group of particles from the Brownian motion of a single particle (Schweitzer 2003). Such models provide ways to describe many kinds of collective behavior, ranging from molecules to the cellular processes that create patterns in the course of development to the movements of ants or migrating wildebeest. Of course, demonstrating that two different processes can be described by the same model does not demonstrate that the processes are alike. Any process can be described by many different models.

The Ecology of Collective Behavior

To learn whether there are general laws of collective behavior that apply across different species and systems, we will have to ask new questions about the ecology of collective behavior and how it evolves. The population geneticist Theodosius Dobzhansky once pointed out that nothing in biology makes sense except in the light of evolution. A corollary is that nothing in evolution makes sense except in the light of ecology. Evolution occurs as a result of the relation between a phenotype and its environment. The next step in the study of collective behavior in animals is to consider the ecology of collective behavior and how it evolves.

Collective behavior is the result of evolutionary processes that shape behavior to modify and respond to environmental conditions (Gordon 2014). Investigating how these algorithms evolve can show how diverse forms of collective behavior arise from their function in diverse environments. Wildebeest migrate to get into an environment with grass to eat. Bees swarm when one nest is too small for the hive, making the old one large enough for the bees that stay behind and putting the new swarm in an appropriate place. Schools of fish turn, and bird flocks move, in response to predators. Schools of dolphins chase their prey, even working together to herd fish into fishermen's nets in return for food (Pryor and Lindbergh 1990).

We can consider the evolution of collective behavior in the same way we think about the evolution of other traits. Rather than asking what individuals have to sacrifice to participate in group activities, we can ask how the collective behavior influences the survival and reproduction of the participants and of the entire system.

To learn about the evolution of collective behavior we must begin to consider variation, the starting point for natural selection. To consider the evolution of individual behavior, we look for differences among individuals. In the same way, to consider the evolution of collective behavior we must look for variation among collectives. Although we speak of the algorithm that a school of fish or an ant colony or a human group uses to accomplish some task collectively, in fact every school, colony, or group behaves differently. Natural selection can occur when this variation is inherited and has some effect on the reproduction of more fish, ants, or people.

Ant colonies are ideal systems in which to consider the evolution of collective behavior (Gordon 2010) because the colonies are the reproductive individuals. A colony consists of one or more female reproductives, or queens, who produce the workers (all sterile females) and also produce reproductive males and daughter queens. Colonies reproduce when reproductive males and daughter queens from different colonies mate and the newly mated queens then go on to found new colonies. The ants in the colony work together to produce reproductives, which mate with the reproductives of other colonies and found new colonies. Thus colonies produce offspring colonies as a result of their collective behavior. Natural selection can shape the way individual ants work together in colonies to produce more colonies. This evolutionary process starts with variation among colonies.

Collective behavior is sometimes equated with cooperative behavior, but these are not the same thing. To identify behavior as collective is to focus on the outcomes at the level of a group of participants. There are many examples of collective behavior in which the participants are not sentient individuals—for example, brain functions produced by the collective behavior of neurons. There are other examples of collective behavior in which the outcomes occur regardless of the motivations and intentions of the participants—a familiar example is a traffic jam. Thus, animals that engage in collective behavior are not necessarily acting according to an intention or motivation to work together.

Cooperation is often defined in terms of costs and benefits: the cooperating individual sacrifices to be part of the group. This compares participation in a group with an imaginary earlier state in which the individual functions on its own. An example of this view comes from early work on social insects. In a colony of social insects, many individuals do not reproduce, but participate collectively in activities that allow other individuals to reproduce. A colony reproduces by producing reproductives that go on to found new colonies. It was suggested that this system could not evolve unless all the individuals shared whatever genes might be associated with worker sterility; otherwise such genes could not persist. Since then, genetic studies have shown that relatedness among social insect workers is not as high as was once supposed, indicating that, however the system originated, it is not currently maintained through kin selection. More important, it is misleading to frame worker sterility as a sacrifice by individuals (Gordon 2012a). Reproduction as a colony entails a sacrifice on the part of individual social insect workers only in the sense that reproduction as a human entails a sacrifice on the part of somatic cells such as those in liver and bone.

Ant colonies vary in behavior. For example, colonies of harvester ants differ in the regulation of behavior (Gordon et al. 2011). Some colonies, year after year, are less likely to forage in poor conditions, when low humidity means a high cost in water loss and when little food is available. This difference arises from differences among colonies in the collective behavior that regulates foraging using interactions between returning foragers bringing food into the nest and outgoing foragers.

Harvester ants foraging in the desert lose water when walking around in the hot sun, and they get their water from the fats in the seeds they eat. Thus the ants must spend water to get water. Faced with the high operating costs of the network that regulates foraging, evolution has opted for positive feedback, so that the system keeps running unless something stops it. Harvester ants use the rate at which ants are returning with food as the source of positive feedback. Each ant returns from its foraging trip, puts down its seed, and waits inside the nest. It does not go out until it meets returning foragers with seeds at a high enough rate. The greater the amount of food that is available, the more rapidly ants will come in with seeds. When food is scarce, ants take a long time to find food and return less frequently to the nest. Few of the foragers in the nest then go out, so little water is lost.

The Transmission Control Protocol (TCP), which manages traffic congestion on the Internet, uses a similar algorithm based on positive feedback (Prabhakar et al. 2012). Each time a data packet succeeds in arriving at a node, an acknowledgment is sent back to the source. The acknowledgment stimulates the transmission of the next data packet. The cost of sending data when there is insufficient bandwidth is high, because the data might be lost. Using this system, data transmission regulates itself to match the available bandwidth.

Variation among ant colonies in collective behavior is due to variation among colonies in how their individual ants respond to interactions (Pinter-Wollman et al. 2013). We are currently using models from neuroscience to describe exactly how ants differ in the ways they assess their rate of interaction, by analogy to the way that neurons act as "leaky integrators" to assess their rate of electrical stimulation (Goldman et al. 2009).

Variation among colonies in the regulation of foraging appears to be inherited by daughter queens from mother queens. A colony lives for 20–30 years, as long as its founding queen survives. All the ants in a colony are the offspring of the founding queen. However, the worker ants live only a year. Year-to-year trends in the collective behavior of the colony thus appear to be inherited by successive cohorts of workers from their mother, the colony's queen. Moreover, it seems that parent and offspring colonies are similar in their sensitivity to current conditions and in their collective decisions about when conditions are poor enough to deter the ants from foraging (Gordon 2013).

Recent work on harvester ants (Gordon 2013) shows that natural selection shapes collective behavior so as to conserve water. The study site in Arizona is subject to the severe and deepening drought throughout the southwestern US in the past 10–15 years. Colonies that are less likely to forage in poor conditions, when water loss is high and the return in food is low, tend to have more offspring colonies. The conditions for natural selection are met: there is variation among colonies in the collective regulation of foraging, this behavior appears to be heritable (and we are currently investigating this), and particular forms of collective behavior lead to higher reproductive success. The collective intelligence of the colony is currently evolving.

Ants are an extremely diverse group of more than 12,000 species that occupy every conceivable habitat on Earth. The diversity of environmental constraints has led to diverse tactics in the collective regulation of colony activity (Gordon 2014). For example, whereas some species search for and retrieve food from a central nest, others circulate along highway systems of permanent trails and search for and retrieve food from the trails (Flanagan et al. 2013, Gordon 2012b).

Similar environmental constraints may lead to the evolution of similar forms of collective behavior in different systems (Gordon 2014). Collective behavior allows animal groups to respond to environmental challenges such as operating costs, the distribution of resources, and the stability of the environment. To examine how collective behavior evolves, we have to investigate whether we see patterns in the response to particular environmental constraints.

Empirical studies in many species are needed to learn how environmental conditions are shaping the evolution of collective behavior. A quantitative description of why a process is effective, or a simulation that selects for that process, helps us to understand how that process works. But to understand its evolution, we need to know its ecological consequences, what problems it solves in a particular environment, and how it is shaped by, and how it influences, changing conditions.

Outlining hypotheses about the fit between collective behavior and its environment can guide the investigation of collective behavior. For example, we now know enough about physiology that we expect animals that live in hot places to have adaptations for heat exchange. In the same way, we can expect the algorithm that dictates collective organization in particular conditions to be tuned to the constraints of those conditions. With respect to the workings of collective biological systems, we are like the European naturalists of the early nineteenth century, agog in the Amazon. We are searching for general trends amid enormous diversity and complexity. A framework for the match between process and environmental conditions can provide predictions that guide the investigation of new systems.

To study the evolution of collective behavior, we have to consider variation among groups in how the algorithm that regulates behavior is deployed, and whether such variation is associated with differences in the ecological success of the collective. Evolution does not always produce the optimal state, and often what is optimal is not apparent. Genetic algorithms can show how evolution could happen, but the selective pressure is created by the programmer, not by nature.

We would like to know whether there are general principles of collective behavior that apply across systems. A first step is to classify the algorithms used, but that is only a beginning. The next step is to consider how collective behavior evolves in natural systems (Gordon 2014). This may reveal trends in the kinds of algorithms used in particular environments. To study the evolution of collective behavior empirically, we have to employ the same methods used to study the evolutionary ecology of individual traits. How do groups vary in their use of a collective algorithm? How does this variation matter to the success of the group? Investigations of how collective behavior functions in its environment are needed.

References

Alon, U. 2007. *An Introduction to Systems Biology: Design Principles of Biological Circuits.* Chapman & Hall/CRC.

Bazazi, S., F. Bartumeus, J. J. Hale, A. J. Holmin, and I. D. Couzin. 2012. Intermittent motion in desert locusts: Behavioral complexity in simple environments. *PLoS Computational Biology* 8 (5): e1002498.

Couzin, I. D., J. Krause, N. R. Franks, and S. A. Levin. 2005. Effective leadership and decision-making in animal groups on the move. *Nature* 433: 513–516.

Deneubourg, J. L., S. Aron, S. Goss, J. M. Pasteels, and G. Duerinck. 1986. Random behavior, amplification processes, and number of participants: How they contribute to the foraging properties of ants. *Physica D* 22: 176–186.

Doyle, J. C., and M. Csete. 2011. Architecture, constraints, and behavior. *Proceedings of the National Academy of Sciences of the United States of America* 108 (Suppl. 3): 15624–15630.

Flanagan, T. P., N. M. Pinter-Wollman, M. E. Moses, and D. M. Gordon. 2013. Fast and flexible: Argentine ants recruit from nearby trails. *PLoS ONE* 8 (8): e70888. doi:10.1371/journal.pone.0070888.

Goldman, M. S., A. Compte, and X. J. Wang. 2009. Neural integrator models. In *Encyclopedia of Neuroscience*, volume 6. Academic Press.

Gordon, D. M. 2010. *Ant Encounters: Interaction Networks and Colony Behavior*. Princeton University Press.

Gordon, D. M. 2012a. What we don't know about the evolution of cooperation in animals. In *Cooperation and Its Evolution*, ed. K. Sterelny et al. MIT Press.

Gordon, D. M. 2012b. The dynamics of foraging trails in the tropical arboreal ant *Cephalotes goniodontus*. *PLoS ONE* 7: e50472.

Gordon, D. M. 2013. The rewards of restraint in the collective regulation of foraging by harvester ant colonies. *Nature* 498: 91–93. doi:10.1038/nature12137.

Gordon, D. M. 2014. The ecology of collective behavior. *PLoS Biology* 12 (3): e1001805. doi:10.1371/journal.pbio.1001805.

Gordon, D. M., A. Guetz, M. J. Greene, and S. Holmes. 2011. Colony variation in the collective regulation of foraging by harvester ants. *Behavioral Ecology* 22: 429–435.

Handegard, N. O., S. Leblanc, K. Boswell, D. Tjostheim, and I. D. Couzin. 2012. The dynamics of coordinated group hunting and collective information-transfer among schooling prey. *Current Biology* 22 (13): 1213–1217.

Nicolis, G., and I. Prigogine. 1977. *Self-Organization in Non-Equilibrium Systems*. Wiley.

Pinter-Wollman, N., A. Bala, A. Merrell, J. Queirolo, M. C. Stumpe, S. Holmes, and D. M. Gordon. 2013. Harvester ants use interactions to regulate forager activation and availability. *Animal Behaviour* 86 (1): 197–207.

Prabhakar, B., K. N. Dektar, and D. M. Gordon. 2012. The regulation of ant colony foraging activity without spatial information. *PLoS Computational Biology* 8 (8): e1002670. doi:10.1371/journal.pcbi.1002670.

Pryor, K., and J. Lindbergh. 1990. A dolphin-human fishing cooperative in Brazil. *Marine Mammal Science* 6: 77–82. doi:10.1111/j.1748-7692.1990.tb00228.x.

Schweitzer, F. 2003. *Brownian Agents and Active Particles: Collective Dynamics in the Natural and Social Sciences*. Springer.

Sumpter, D. J. T. 2010. *Collective Animal Behavior*. Princeton University Press.

Human–Computer Interaction

Editors' Introduction

The field of human–computer interaction (HCI) contributes to collective intelligence by designing and studying the digital channels that give rise to it. For example, how is power distributed in Wikipedia? How might changing Wikipedia's processes or interface alter those behaviors? HCI research seeks to answer these kinds of questions by focusing on how design decisions at the individual level lead to emergent effects at the collective level. Rather than studying just the human or the design, HCI research considers the system and its social dynamics together.

HCI research has one foot in design, aiming to create new kinds of collectives. For example, one thread of research spawned a series of online games. These games have side effect of producing training data so that computers can solve difficult problems such as image labeling or protein folding. Likewise, crowd-powered systems embed paid crowd workers alongside traditional software interfaces so that Microsoft Word users can have access to on-demand text editing, blind users can take photos and ask questions about their surroundings, or lectures can be transcribed in real time. This design orientation is shared by the chapter on artificial intelligence and the chapter on organizational behavior.

HCI's other foot is firmly in social science, pursuing observational and experimental studies of existing collectives. HCI researchers blend a background in the methods of social science with the tools of computer science to build theories using the issues in online communities or the behavioral data that the communities generate. The methods of their studies range from ethnomethodological or qualitative investigations to online field experiments and data mining. HCI researchers attempt to answer questions such as "Why do online communities succeed or fail?" and "Is there a filter bubble that prevents different sides of a debate from engaging with one another?" In this sense, HCI draws on many of the methods discussed in the chapter on cognitive psychology and in the chapter on law and other disciplines.

Unlike many other disciplines, HCI does not take the technological infrastructure as fixed. It actively develops new ways for people to come together, then seeks to understand what works and what doesn't. HCI inherits its epistemology more from a blend of design, psychology, and anthropology. It is an inherently interdisciplinary enterprise.

Social computing, crowdsourcing and collective intelligence together make up just one community of HCI researchers. For example, there is an active community working on ubiquitous computing technologies, aiming to make computing such a natural part of our physical environment that it disappears from our active attention. Other communities, including those working in critical design and design theory, focus on the process of creating new designs. Still other communities care about collaboration in small workgroups, or design for developing regions, or interaction techniques and hardware at scales both large and small.

In the chapter that follows, Bigham, Bernstein, and Adar use crowdsourcing to examine the intersection of human–computer interaction and collective intelligence. It starts with directed crowdsourcing, in which a leader such as an end user is guiding the process that the crowd follows. It then widens to consider collaborative crowdsourcing, in which the group self-determines its actions. The chapter closes with data-driven approaches ("passive crowdsourcing"), focusing on behavioral traces that users leave behind to be mined.

We recommend pairing the chapter by Bigham, Bernstein, and Adar with the chapter in which Benkler, Shaw, and Hill discuss the surprising success of volunteer online movements such as Wikipedia and open-source software. Both chapters discuss Wikipedia. Whereas Benkler et al. present a deeper examination of the encyclopedia, Bigham et al. situate it in the movements of crowdsourcing, social computing and data mining.

Recommended Readings

Bill Buxton. 2010. *Sketching User Experiences: Getting the Design Right and the Right Design.* Morgan Kaufmann.

In this book, Bill Buxton captures many of the modern perspectives and techniques behind the iterative, user-centered design process. This perspective undergirds much of

HCI today: a deep appreciation for the human element of technology, combined with a technologist's eye for creating new solutions to help people achieve their goals.

Luis von Ahn and Laura Dabbish. 2004. Labeling images with a computer game. In *Proceedings of the SIGCHI Conference on Human Factors in Computing Systems*. ACM.

Luis von Ahn's ESP Game took the Web by storm, and many players spent hours playing that image-labeling game with anonymous Internet partners. By trying to guess what the other player would say about an image, thousands of players collectively generated a large annotated dataset of images that could be used to make the Web more searchable.

Michael S. Bernstein, Greg Little, Robert C. Miller, Björn Hartmann, Mark S. Ackerman, David R. Karger, David Crowell, and Katrina Panovich. 2010. Soylent: A word processor with a crowd inside. In *Proceedings of the 23nd Annual ACM Symposium on User Interface Software and Technology*. ACM.

Crowd-powered systems aim to advance crowdsourcing from a batch platform to an interactive platform that operates in real time. The research discussed in this paper develops computational techniques that decompose complex tasks into simpler, verifiable steps, return results in seconds, and open up a new design space of interactive systems. Soylent uses paid micro-contributions to help writers to edit and correct text. Using Soylent is like having an entire editorial staff available as you write. For example, crowds notice wordy passages and offer alternatives that the user may not have considered.

Aniket Kittur, Boris Smus, and Robert E. Kraut. 2011. Crowdforge: Crowdsourcing complex work. In *Proceedings of the 24th Annual ACM Symposium on User Interface Software and Technology*. ACM.

Directed crowdsourcing has traditionally been applied to situations in which work is straightforward and quick. But could it apply in much more complex scenarios, such as trying to write an encyclopedia article or a blog post? The impact of the Crowdforge paper was to set such complex goals as a target for crowdsourcing research. It proposed a computational model for splitting this work and recombining it.

Daniela Retelny, Sébastien Robaszkiewicz, Alexandra To, Walter Lasecki, Jay Patel, Negar Rahmati, Tulsee Doshi, Melissa Valentine, and Michael S. Bernstein. 2014. Expert crowd-sourcing with flash teams. In *Proceedings of the 27th Annual ACM Symposium on User Interface Software and Technology*. ACM.

Flash teams is an effort to envision a future in which paid expert workers can assemble, on demand, into computationally managed micro-organizations. Experts in programming, art, music, writing, and other fields have been increasingly populating online labor marketplaces such as Upwork, Freelancer, and TopCoder. Flash teams draw on these labor markets to advance a vision of expert crowd work that accomplishes complex, interdependent goals such as engineering and design. The teams consist of sequences of linked modular tasks and handoffs that can be managed computationally.

Firas Khatib, Seth Cooper, Michael D. Tyka, Kefan Xu, Ilya Makedon, Zoran Popović, David Baker, and Foldit players. 2011. Algorithm discovery by protein folding game players. *Proceedings of the National Academy of Sciences* 108 (47): 18949–18953.

Could human game players solve problems of protein folding that have evaded scientists for years? The creators of Foldit launched an online game to answer just this question, and the answer was a resounding Yes. Since its launch, Foldit and its players have published several papers in top-tier scientific journals.

Human–Computer Interaction and Collective Intelligence

Jeffrey P. Bigham, Michael S. Bernstein, and Eytan Adar

The field of Human–Computer Interaction (HCI) works to understand and to design interactions between people and machines. Increasingly, human collectives are using technology to gather together and coordinate. This mediation occurs through volunteer and interest-based communities on the Web, through paid online marketplaces, and through mobile devices.

The lessons of HCI can therefore be brought to bear on different aspects of collective intelligence. On the one hand, the people in the collective (the crowd) will contribute only if there are proper incentives and if the interface guides them in usable and meaningful ways. On the other, those interested in leveraging the collective need usable ways of coordinating, making sense of, and extracting value from the collective work that is being done, often on their behalf. Ultimately, collective intelligence involves the co-design of technical infrastructure and human–human interaction: a socio-technical system.

In crowdsourcing, we might differentiate between two broad classes of users: requesters and crowd members. The requesters are the individuals (or the group) for whom work is done or who take responsibility for aggregating the work done by the collective. The crowd member (or crowd worker) is one of many people who contribute. Although we often use the word "worker," crowd workers do not have to be (and often aren't) contributing as part of what we might consider standard work. They may or may be paid, they may work only for short periods of time or spend days contributing to a project they care about, and they may work in such a way that each individual's contribution may be difficult to discern from the collective final output.

HCI has a long history of studying not only the interaction between individuals with technology but also the interaction of groups with or mediated by technology. For example, computer-supported cooperative work (CSCW) investigates how to enable groups to accomplish tasks together using a shared or distributed computer interfaces, either at the same time or asynchronously. Current crowdsourcing research alters some of the standard assumptions about the size, composition, and stability of these groups, but the fundamental approaches remain the same. For instance, workers drawn from the crowd may be less reliable than groups of employees working on a shared task, and group membership in the crowd may change more quickly.

There are three main vectors of study for HCI and collective intelligence. The first is *directed crowdsourcing*, in which an individual attempts to recruit and guide a large set of people to help accomplish a goal. The second is *collaborative crowdsourcing*, in which a group gathers on the basis of shared interest and its members self-determine their organization and work. The third vector is *passive crowdsourcing*, in which the crowd or collective may never meet or coordinate, but it is still possible to mine their collective behavior patterns for information. In this chapter we cover these vectors in turn. We conclude with a list of challenges for researchers in HCI—challenges related to crowdsourcing and collective intelligence.

Directed Crowdsourcing

The term "directed crowdsourcing" describes systems in which an algorithm or a person directs workers to pursue a specific goal. For example, a requester might seek to gather a crowd to tag images with labels or to translate a poem. Typically, this involves a single requester taking a strong hand in designing the process for the rest of the crowd to follow.

This section gives an overview of work in directed crowdsourcing, including considerations to be made when deciding whether a particular problem is amenable to directed crowdsourcing, how to design tasks, and how to recruit and involve workers.

Workers in directed crowdsourcing generally complete tasks that the requester asks to be completed. Why would they perform the task? Sometimes the goals of requesters and workers are aligned, as is the case in much of the crowdsourcing work being done in citizen science. For instance, the crowd cares about a cause, such as tracking and counting birds (Louv et al. 2012), and the requester's direction is aimed primarily at coordinating and synthesizing the work of willing volunteers.

In other situations, crowd workers may not share the same goal as the requester. In this case, one challenge is to design incentives that encourage them to participate. This can be tricky because workers may have different reasons for participating in crowd work: for example, money, fun, or learning.

Several systems have "gamified" (that is, introduced elements of games into) crowd tasks to incentivize workers by making the tasks more enjoyable. For instance, the ESP Game is a social game that encourages players to label images by pairing the players with partners who are also trying to label the same images (von Ahn and Dabbish 2004). Likewise, Foldit players found stable protein configurations that had eluded scientists for a decade (Khatib et al. 2011). Gamification can take years, and converting tasks to games that people will want to play may require a great deal of insight. It can also be difficult to attract players to a game, and a game's popularity may fluctuate. Some of these games have attracted numerous players; many others have not.

Another option is to pay crowd workers. Paid crowdsourcing differs from traditional contract labor in that it often occurs at very small timescales (for similarly small increments of money), and often interaction with the worker can be fully or partially programmatic. This form of crowdsourcing often takes place in online marketplaces such as Amazon Mechanical Turk and Upwork. In paid crowdsourcing, a worker's incentive is ostensibly money, although money is not the only motivator even in paid marketplaces (Antin and Shaw 2012). Money may affect but cannot be reliably used to improve desirable features of the contributions, e.g. the quality or the timeliness of the work (Mason and Watts 2010). Because of the ease by which paid workers can be recruited, paid crowdsourcing, especially Amazon Mechanical Turk, is a popular prototyping platform for research and product in crowdsourcing.

An alternative approach is to collect crowdsourcing as a result (or a by-product) of something else the user (or worker) wanted to do. For instance, reCAPTCHA is popular service that attempts to differentiate people from machines on the Web by presenting a puzzle made up of blurry text that one must decipher in order to prove that one is human (von Ahn et al. 2008). As opposed to other CAPTCHAs, reCAPTCHA has a secondary goal of converting poorly scanned books to digital text. reCAPTCHA presents two strings of text, one that it knows the answer to and one that it does not. By typing the one it knows the answer to, the user authenticates himself. By typing the one it does not know, the user contributes to the goal of digitizing books. DuoLingo uses a similar approach to translate documents on the Web into new languages as a by-product of users' learning foreign languages (Hacker and von Ahn 2012).

Quality and Workflows

When a requester asks for people to help with a task, they will often do exactly what is asked but not quite what was desired. Quality-control mechanisms and workflows can help ensure better results. Interestingly, however, the quality of outputs from paid crowdsourcing is not a clear function of the price paid. For instance, paying more has been found to increase the quantity but not necessarily the quality of the work that is done (Mason and Watts 2010). The usability of the task at hand can affect how well workers perform at the task. Human–computer interaction offers crowdsourcing methods for engineering tasks that crowd workers are likely to be able to do well with little training. Because crowd workers are often assumed to be new to any particular task, designing to optimize learnability is important; other usability dimensions, such as efficiency or memorability, may be less so.

Crowdsourcing tasks are often decomposed into small bits of work called microtasks. As a result, workers may not understand how their contribution fits into a broader goal, and this may affect the quality of their work. One way to compensate for the variable quality of the work received and to combine the small efforts of many workers is to use a workflow, also sometimes called a crowd algorithm. Some common workflows are iterative improvement (Little et al. 2009), parallel work followed by a vote (ibid.), map-reduce (Kittur et al. 2011), find-fix-verify (Bernstein et al. 2010a), and crowd clustering (Chilton et al. 2013). Good workflows help to achieve results that approach the performance of the average worker in the crowd, and sometimes can help achieve the "wisdom of the crowd" effect (in which the group performs better than any individual). Good workflows also allow a large task to be completed consistently even if each worker works on the task for only a short time.

For example, Soylent (see figure 4.1) is a Microsoft Word plug-in that allows a crowd to help edit documents by, for instance, fixing spelling or grammar, or shortening the document without changing its meaning. It introduced the Find-Fix-Verify workflow, which proceeds in three steps:

1. Workers find areas in the document can could be appropriate for improvement.
2. A second set of workers propose candidate changes (fixes).
3. A third set of workers verify that the candidate changes would indeed be good changes to make.

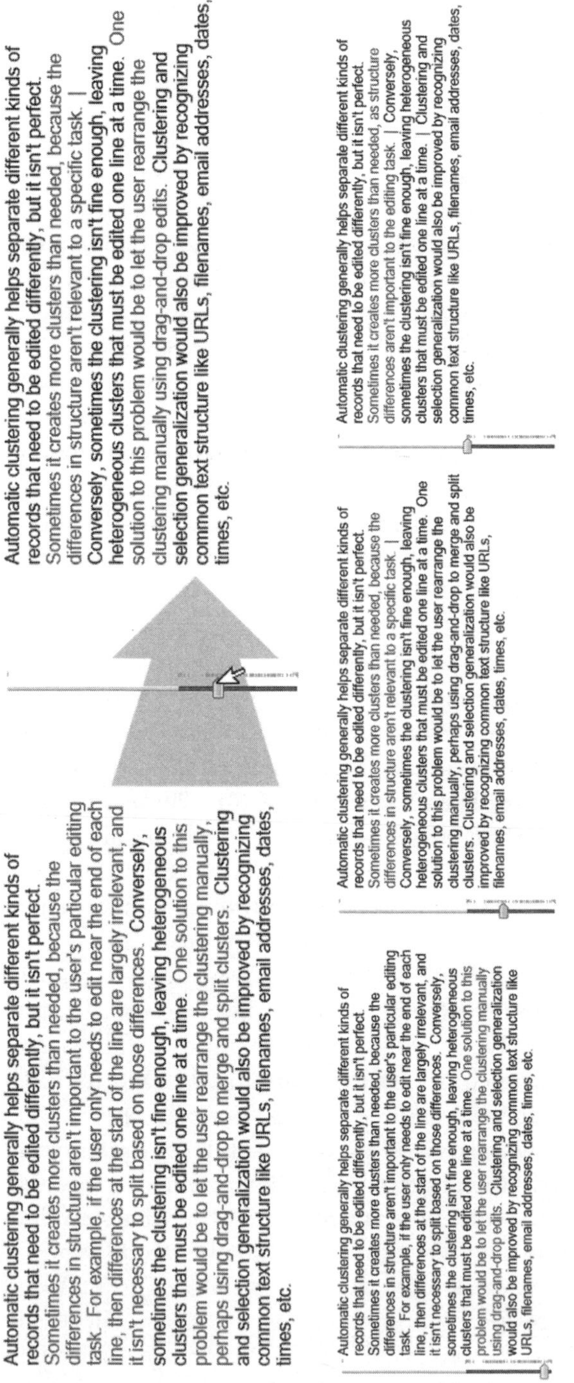

Automatic clustering generally helps separate different kinds of records that need to be edited differently, but it isn't perfect. Sometimes it creates more clusters than needed, because the differences in structure aren't important to the user's particular editing task. For example, if the user only needs to edit near the end of each line, then differences at the start of the line are largely irrelevant, and it isn't necessary to split based on those differences. Conversely, sometimes the clustering isn't fine enough, leaving heterogeneous clusters that must be edited one line at a time. One solution to this problem would be to let the user rearrange the clustering manually, perhaps using drag-and-drop to merge and split clusters. Clustering and selection generalization would also be improved by recognizing common text structure like URLs, filenames, email addresses, dates, times, etc.

Automatic clustering generally helps separate different kinds of records that need to be edited differently, but it isn't perfect. Sometimes it creates more clusters than needed, because the differences in structure aren't important to the user's particular editing task. For example, if the user only needs to edit near the end of each line, then differences at the start of the line are largely irrelevant, and it isn't necessary to split based on those differences. Conversely, sometimes the clustering isn't fine enough, leaving heterogeneous clusters that must be edited one line at a time. One solution to this problem would be to let the user rearrange the clustering manually using drag-and-drop edits. Clustering and selection generalization would also be improved by recognizing common text structure like URLs, filenames, email addresses, dates, times, etc.

Automatic clustering generally helps separate different kinds of records that need to be edited differently, but it isn't perfect. Sometimes it creates more clusters than needed, because the differences in structure aren't relevant to a specific task. | Conversely, sometimes the clustering isn't fine enough, leaving heterogeneous clusters that must be edited one line at a time. One solution to this problem would be to let the user rearrange the clustering manually, perhaps using drag-and-drop to merge and split clusters. Clustering and selection generalization would also be improved by recognizing common text structure like URLs, filenames, email addresses, dates, times, etc.

Automatic clustering generally helps separate different kinds of records that need to be edited differently, but it isn't perfect. Sometimes it creates more clusters than needed, as structure differences aren't important to the editing task. | Conversely, sometimes the clustering isn't fine enough, leaving heterogeneous clusters that must be edited one line at a time. | Clustering and selection generalization would also be improved by recognizing common text structure like URLs, filenames, email addresses, dates, times, etc.

Figure 4.1
Soylent reaches out to paid crowds for editorial help within a word processor. It allows users to adjust the length of a paragraph via a slider. Lighter text (red in the original figure) indicates locations where the crowd has provided a rewrite or cut. Tick marks on the slider represent possible lengths.

The FFV workflow has a number of benefits. First, as had been observed previously, workers tended to make the smallest acceptable change. For instance, if they were asked to directly fix the document or to make it smaller, workers would find a single change to make. The "find" step encourages multiple workers to find many things to fix in the document (or in their assigned chunk of the document). The "fix" and "verify" steps are then scoped to that particular location. That was observed to result in more problems' being found and fixed.

Workflows can often get complex, requiring many layers of both human and machine interaction. For instance, PlateMate combines several crowd-powered steps with machine-powered steps into a complex workflow that is able to match the performance of expert dieticians in determining the nutritional content of a plate of food (Noronha et al. 2011). For a new problem that one wants to solve with crowdsourcing, coming up with an appropriate workflow that allows crowd workers to contribute toward the end goal can be a challenge.

As crowdsourcing broadens from amateur microtasks to goals involving groups of interdependent experts, the nature of these work-flows may change. Flash Teams offer one vision, in which computation acts as a coordinating agent to draw together diverse experts from platforms such as Upwork (Retelny et al. 2014). These kinds of approaches have already made possible the crowdsourcing of a broad class of goals, including design prototyping, course development, and film animation, in half the work time that traditional self-managed teams require.

Interactive Crowd-Powered Systems

Traditional workflows can be quite time consuming, as each iteration requires crowd workers to be recruited and to perform their work. Nearly real-time workflows use time as a constraint and often have people work in parallel and then have either an automatic or a machine process make sense of the work as it is produced. The first step is to pre-recruit a group of workers who are then available to do work at interactive speeds once the work to be done is available. The SeaWeed system pre-recruited a group of workers who would then collectively play economics games (Chilton et al. 2009). VizWiz pre-recruited workers and had them answer old questions until a new question came in for them (Bigham et al. 2010). Adrenaline used a retainer pool to recruit a group of workers and then showed that this group could be called back quickly (Bernstein et al. 2011). Workers recruited into the retainer pool are paid a small bonus to be part of the pool, and collect these earnings if they respond quickly enough when asked. Turkomatic recruits workers and then lets them be programmatically sent to a task as a group (Kulkarni et al. 2012).

There is also value in getting workers to work together synchronously. One reason to do this is to build real-time systems that are able to compensate for common problems in the crowd—namely, that workers sometimes perform poorly and sometimes leave the task for something else to do. For instance, the Legion system puts crowd workers in control of a desktop interface by having them all contribute keyboard and mouse commands (Lasecki et al. 2011). The crowd worker who for a given time interval is most similar to the others is elected a leader and given full control, thus balancing the wisdom of the crowds with a real-time constraint. This system was used across a variety of desktop computing applications and even to control a wireless robot. Adrenaline uses a similar concept to quickly refine frames of interest in a video and then eventually pick a high-quality frame from a digital video, thus creating a real-time crowd-powered camera.

Another reason to work as a group is to accomplish a goal that no worker could accomplish alone. The Scribe system allows a group of workers to collectively type part of what they hear a speaker say in real time (Lasecki et al. 2012). An automated process then stitches the pieces back together using a variant of Multiple Sequence Alignment (Naim et al. 2013). No worker is able to keep up with natural speaking rates alone, but collectively they can do so by using this approach. Employing a group allows the task to be made easier in ways that would not be possible if a single person were responsible for typing everything. Most obviously, each worker has to type only part of what he hears. More interestingly, the task of each worker in a group can be made even easier. The audio of the portion of speech the workers is expected to type can be algorithmically slowed down, which allows the worker to keep up more easily (Lasecki et al. 2013). The remainder of the audio is then sped up so that the worker can keep context. Overall, this increases recall and precision, and reduces latency.

Programming with the Crowd

Crowd-powered systems behave differently than completely automated systems, and a number of programming environments have

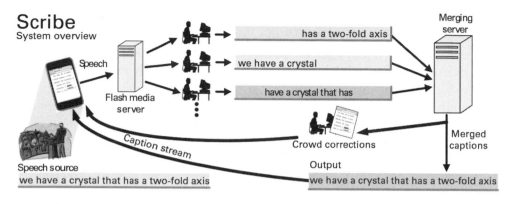

Figure 4.2
Legion:Scribe allows users to caption audio on their mobile device. The audio is sent to multiple amateur captionists who use the Legion:Scribe Web-based interface to caption as much of the audio as they can in real time. These partial captions are sent to the Legion server to be merged into a final output stream, which is then forwarded back to the user's mobile device. Crowd workers are optionally recruited to edit the captions after they have been merged.

been constructed to assist in designing and engineering them. For instance, crowd workers are often slow and expensive, so TurKit allows programs to reuse results from tasks that have executed but have not finished, employing a "crash and run" paradigm that allows for easier programming (Little et al. 2009). Both VizWiz and Soylent were programmed using TurKit as a scaffold. AskSheet embeds crowd work into a spreadsheet and helps to limit the steps that crowd workers have to take in order to make decisions (Quinn and Bederson 2014).

Jabberwocky exposes workflows as programming language primitives and supports operation on top of a number of different kinds of crowds, including Mechanical Turk and social sources such as Facebook (Ahmad et al. 2011). One of the workflows it makes easily available is a crowdsourcing equivalent of Map Reduce called Man Reduce. Jabberworky builds on CrowdForge's approach by having work automatically divided up in the Map step for each of a number of workers to each complete and then recombined in the reduce step (Kittur et al. 2011). One example of this is writing an essay by assigning different paragraphs to different workers and then having a reduce step in which those paragraphs are recombined.

When crowd-powered systems do not behave as expected, it can be difficult to figure out why. Some systems have been developed to allow for the equivalent of debugging. For instance, CrowdScape records low-level features of how crowd workers perform their task and then allows requesters to easily visualize the recordings (Rzeszotarski and Kittur 2012). This can help requesters to identify confusing aspects of the work, to understand where improvements are most needed (e.g., in code or the crowd tasks), and to understand performance, even on subjective tasks. The scrolling, key presses, mouse clicks, and other actions that collectively define the "fingerprint" of a task can be useful in understanding how the work was done. If a user was asked to read a long passage and then answer a question about it, we might assume that the work was not done well if the user scrolled quickly past the text and answered immediately.

Drawbacks and Ethics of Microtasking

The field of human–computer interaction is acutely aware of how the sociotechnical systems that are created may affect the future of crowd work. Crowds bear the potential of mass action and people power. Yet, as Irani and Silberman (2013) have argued, Amazon Mechanical Turk's design directs this collective power into reliable, steadily humming computational infrastructure. This infrastructure is designed to keep questions of ethical labor relations or worker variation out of requesters' (employers') sight (Salehi et al. 2015). This may lead to undesirable consequences for crowdsourcing, such as divorcing workers from the tasks they work on and reducing the value assigned to expertise. Microtasks remain popular despite their drawbacks because they are accessible and because they often require low effort for requesters to create.

In the past, crowd work sometimes was viewed as a source of very low-cost labor. Because such labor sometimes provides low-quality input, techniques have to be derived to compensate for it. One of the goals of HCI research in crowdsourcing is to demonstrate the potential for a brighter future for crowd work in which workers are able to accomplish together something that they could not have accomplished on their own.

It may be tempting in crowd work to treat workers as program code. Some have recognized that this prevents many of the benefits of crowd workers' being human from being realized. For instance, once crowd workers learn a new task, they are likely to be better (faster, more accurate) at it than workers newly arriving to the task. As a result, it is advantageous to requesters to try to retain workers over time, and to have them working on similar tasks, so that the workers can become more skilled and more efficient. This is also advantageous to the workers, who can complete more tasks and earn more.

Concerns about labor practices have led to work exploring current demographics of workers and work that explicitly considers how to improve working conditions. In a paper titled "The future of crowd work," Kittur et al. (2013) make a number of suggestions for improving crowd work, including allowing workers to learn and acquire skills from participating in such work. It may be possible to design socio-technical systems that enable workers to self-organize, though these efforts have threatened to either stall or flare into acrimony (Salehi et al. 2015). A common but incorrect notion about Mechanical Turk, for instance, is that workers are mostly anonymous—however, this has since been shown not to be true (Lease et al. 2013). HCI researchers are increasingly taking into account both the advantages of a programmatically available source of human advantage and the essential humanness of the participants.

Collaborative Crowdsourcing

Many of the most famous crowdsourcing results are not directed. Instead, they depend on volunteerism or other non-monetary incentives for participation. For example, volunteer crowds have authored Wikipedia (http://www.wikipedia.org), the largest encyclopedia in history, helped NASA identify craters on Mars (Kanefsky et al. 2001), surveyed satellite photos for images of a missing person (Hellerstein and Tennenhouse 2011), held their own in chess against a world champion (Nalimov et al. 1999), solved open mathematics problems (Cranshaw and Kittur 2011), generated large datasets for object recognition (Russell et al. 2008), collected eyewitness reports during crises and violent government crackdowns (Okolloh 2009), and generated a large database of common-sense information (Singh et al. 2002). Each of these successes relied on the individuals' intrinsic motivation to participate in the task. Intrinsic motivation, unlike extrinsic motivators such as money, means that participants bring their own excitement to the table: for example, via interest in the topic, desire to learn, or drive to experience new things (Ryan and Deci 2000).

Although sociotechnical systems often have leaders, as in Jimmy Wales' leadership of Wikipedia or the main contributors to an open-source project, we refer to these systems as *collaborative crowdsourcing* efforts here. Doing so distinguishes them from the earlier efforts in which a single individual is more directly driving the group's vision and tasks. Often, in collaborative crowdsourcing, the group exercises self-determination and self-management, as a form of meta-work, to plan the group's own future before executing it.

HCI research first seeks to understand these sociotechnical systems. Why do they work? What do they reveal about the human processes behind collective intelligence? How do changes to the design or tools influence those processes? Second, in parallel, HCI research aims to empower these self-directed systems through new designs. These designs may be minor changes that produce large emergent effects—for example, recruiting more users to share movie ratings (Beenen et al. 2004)—or they may be entirely new systems—for example, a community created to capture 3-D models of popular locations through photographs (Tuite et al. 2011).

HCI research engages with themes such as leadership, coordination, and conflict as well as through the type of information provided by the crowd. For each aspect, many decisions will influence the effectiveness and the fit of a particular design for different contexts.

Leadership and decision making

When a group is self-organizing, decision making becomes a pivotal activity. Does the group spend more time debating its course of action than it spends making progress?

Kittur et al. (2007) undertook one the best-known explorations of this question. They obtained a complete history of all Wikipedia edits, then observed the percentage of edits that were producing new knowledge (e.g., edits to article pages) vs. edits that were about coordinating editing activities (e.g., edits to talk pages or policy pages). Over time, the number of article edits decreased from roughly 95 percent to just over 50 percent of the activity on Wikipedia (figure 4.3). This result suggests that as collective intelligence activities grow in scope and mature they may face increased coordination costs.

Leadership faces other challenges. Follow-on work discovered that as the number of editors on an article increases, the article's quality increases only if the editors take direct action via edits rather than spend all their effort debating in Wikipedia's talk pages (Kittur and Kraut 2008). In policy decisions, senior members of the community have a greater-than-average ability to kill a proposal, but no more than an average ability to make a proposal succeed (Keegan and Gergle 2010). In volunteer communities, it may be necessary to pursue

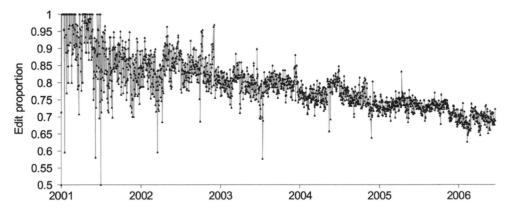

Figure 4.3
Kittur and Kraut observed that as Wikipedia matured, fewer and fewer edits were actually contributing content to the encyclopedia, and more edits were on "talk" and policy pages where meta-work (e.g., coordination) was happening.

strategies of distributed leadership (Luther et al. 2013). These strategies can be more robust to team members' unexpectedly dropping out or failing to deliver promised work.

In terms of design, socializing new leaders is challenging. Whether future leaders are different even starting from their earliest activities (Panciera et al. 2009) or whether they are just like other members and can be scaffolded up through legitimate peripheral participation (Lave and Wenger 1991; Preece and Shneiderman 2009) is a subject of debate. Software tools to help train leaders can make the process more successful (Morgan et al. 2013).

Coordination
Crowds can undertake substantial coordination tasks on demand. In these situations, emergent coordination among crowds takes the place of explicit leadership. Crisis informatics, for example, focuses on coordination in the wake of major disasters such as earthquakes and floods. Groups have adopted social media such as Twitter to promote increased situational awareness during such crises (Vieweg et al. 2010). When official information is scarce and delayed, affected individuals can ask questions and share information; remote individuals can help filter and guide that information so it is most effective (Starbird 2013).

Coordination can be delicate. On one hand, successful scientific coordinations such as the Polymath project demonstrate that loosely guided collaborations can succeed (Cranshaw and Kittur 2011). In the Polymath project, leading mathematicians blogged the process of solving a mathematics problem and recruited ideas and proofs from their readers. On the other hand, distributed voting sites such as Reddit exhibit widespread underprovision of attention (Gilbert 2013)—in other words, the users' attention is so focused on a few items that they often miss highly viral content the first time it is posted. Platforms such as Kickstarter may offer a "living laboratory" (Chi 2009) for studying collective coordination efforts at large scales (Gerber et al. 2012).

Conflict

Most collective intelligence systems will produce internal conflict. Some systems (e.g., systems to support deliberative democracy) are even *designed* to host conflict. The objective of such designs is not to remove or resolve disagreement, but rather to show participants points of agreement and contention.

For example, Reflect (Kriplean et al. 2011b) and ConsiderIt (Kriplean et al. 2011a) are designed to host discussion and debate. In order to do so, they introduce procedural changes into the format of a discussion. For example, Reflect asks each commenter to first summarize the original poster's points. ConsiderIt, which is focused on state election propositions, asks visitors to write well-structured pro and con points rather than leave unstructured comments.

Design may also aim to increase awareness of other perspectives. By visualizing to what extent users' consumption of news is biased, browser plug-ins can encourage them to balance their news perspectives (Munson et al. 2013). Projects such as OpinionSpace (Faridani et al. 2010) and Widescope (Burbank et al. 2011) likewise demonstrate how even people who disagree on binary choices may, in practice, be closer in opinion than they think they are.

Participation

Though leadership, cooperation, and the management of conflict are important in the construction of effective crowd systems, continuous active participation is critical. Without participation in collective intelligence activities, there is no collective, and thus no intelligence. The literature on online communities has devoted considerable energy to studying how to attract and maintain participation. (In this volume, ideas of incentives and motivation are also discussed in the chapter on artificial intelligence and in the chapter on law and other disciplines.)

The GroupLens project has produced some of the most influential research investigating this question. Years ago, GroupLens created a website called MovieLens. An early movie-recommendation service, it attracted a constant stream of volunteers. The researchers then began applying concepts from social psychology to increase participation to MovieLens. For example, they found that calling out the uniqueness of a user's contributions and creating challenging but achievable goals increased the number of movies that users would rate on the site (Beenen et al. 2004).

Other successful approaches include creating competitions between teams (Beenen et al. 2004) or calling out the number of other people who have also contributed (Salganik and Watts 2009). Kraut and Resnick's 2012 book *Building Successful Online Communities* provides thorough references.

Information seeking and organizational intelligence

Human–computer interaction has long focused on user interaction with many kinds of information. Thus, a second set of decision points relate to the kind of information we would like to pull from the crowd. For example, in some situations, this information already exists in the heads of other individuals. Mark Ackerman introduced the idea of actively recruiting and gathering this knowledge through a system called Answer Garden (Ackerman and Malone 1990). Answer Garden was a precursor to such question-and-answer (Q&A) systems as Yahoo, Answers, and Quora. It encouraged members of organizations to create reusable knowledge by asking questions and retaining the answers for the next users who wanted them.

Since Answer Garden, Q&A systems have matured into products such as the social search engine Aardvark (Horowitz and Kamvar 2010). Recent work (e.g., Evans and Chi 2008; Morris et al. 2010) has focused on social search, in which users ask their own social networks (i.e., their friends) for answers. Often this "friendsourcing" approach (Bernstein et al. 2010b) can solve problems that generic crowds cannot.

Exploration and discovery

While focused on a particular problem, crowds generate diverse perspectives and insights; as Eric Raymond put it, "given enough eyeballs, all bugs are shallow." Thus, it isn't surprising that some of the most influential crowdsourcing communities have focused on discovery.

The protein-folding game Foldit is the best-known example of crowd discovery. (See Foldit 2008.) It has attracted nearly 250,000 players, and they uncovered protein-folding configurations that had baffled scientists for years (Cooper et al. 2010). That these findings have been published in a journal as influential as *Nature* suggests something about a crowd's ability to solve difficult and important problems.

Foldit attracted novices, something that is not uncommon when a scientific goal holds intrinsic interest; the Galaxy Zoo project, which labels galaxy images from a star survey (Lintott et al. 2008), is another good example. Cooperative crowdsourcing tools may also allow users to focus deeply on micro-areas of interest; for example, collaborative visualization tools such as sense.us (Heer et al. 2007) and ManyEyes (Viégas et al. 2007) allowed users to share visualizations and collaboratively work to explain interesting trends.

Creativity

Can crowds be creative? Certainly members of a crowd can be. For example, an online community called Scratch allows children to create and remix animations (Resnick et al. 2009). However, it isn't clear that remixing produces higher-quality output (Hill and Monroy-Hernández 2013). In a more mature setting, members of the Newgrounds animation site spend many hours creating collaborative games and animations. Their collaborations are delicate and don't always succeed (Luther and Bruckman 2008). When they do succeed, the onus is on the leader to articulate a clear vision and communicate frequently with participants (Luther et al. 2010).

HCI pursues an understanding of how best to design successful creative collaborations. For example, it may be that structuring collaborative roles to reflect the complementary creative strengths of the crowd and the individual can help. Ensemble is an application for collaborative creative writing in which a leader maintains a high-level vision and articulates creative constraints for the crowd, and the crowd generates text and contributions within those constraints (Kim et al. 2014). In the domain of music, data from the annual February Album Writing Month (FAWM) uncovered how complementary skill sets can be predictive of successful creative collaborations (Settles and Dow 2013).

Collective action

Volunteer crowds can come together to effect change in their world. Early in the days of crowdsourcing, this situation hit home with the academic computer science community when a well-known professor at UC Berkeley named Jim Gray disappeared at sea while flying his plane. The community rallied, quickly hacking together software to search satellite images of the region to find Gray's downed plane (Hellerstein and Tennenhouse 2011).

The rapid collaborative responses to this tragedy was a precursor for research projects that collect and study other collective-action efforts and for the design of new systems to support collective action. As with all collective action problems, getting off the ground can be a challenge. Thus, Catalyst allows individuals to condition their participation on others' interest, so that I might commit to tutoring only if ten people commit to attending my tutoring session (Cheng and Bernstein 2014). We may yet see examples of crowds coming together not just to talk, but to *act*, as Mechanical Turk workers have done on Dynamo (www .wearedynamo.org).

Passive Crowdsourcing

Crowdsourcing is often perceived as requiring a requester to make direct elicitation of human effort. However, the relationship between a requester and a crowd can also be indirect. In *passive crowdsourcing* the crowd produce useful "work product" simply as part of their regular behavior. That is, the work is a side effect of what people were doing ordinarily. Rather than directing the efforts of the crowd as in the active scenarios, the requester is passively monitoring behavioral traces and using them to make decisions.

As a simple example, take a Web search engine that collects user logs of search and click behavior (i.e., which results are clicked after the search). The system observes that when most users search for some concept (say, "fruit trees") they conclude their search session on a particular entry in the search results page (say, the third result). From this the system infers that the third page probably contains the best answer to the query and the result is boosted to the top of the list (Culliss 2000). The crowd here is doing work for the "requester" (they are helping organize search results), but this is simply a side effect of how they would use the system ordinarily—that is, to find results of interest.

The difficulty in this approach is that passive crowdsourcing systems are explicitly designed to avoid interfering with the worker's "ordinary" behavior. This requires effective instrumentation, calibration, and inference that allows the system designer to go from a noisy signal that is at best weakly connected to the desired work product to something useful. For example, the search engine does not directly ask end users to indicate which result is the best; rather, end users *observe* the click behavior and *infer* the best result.

Passive designs are often used to achieve some effect in the original system (e.g., better search results), but the traces can also be used for completely different applications. For example, companies such as AirSage (2014) have utilized the patterns by which cell phones switch from tower to tower as people drive (the "ordinary" behavior that allows cell phones to function) in order to model traffic flow and generate real-time traffic data. In all of these instances, there is no explicit "request" being made to the crowd, but the crowd is nonetheless doing work.

Examples of passive crowd work

The idea of non-reactive measures has a significant history in the sociological literature (Webb et al. 1999), in which researchers identified mechanisms for collecting data without directly asking subjects. The quintessential example is the identification of the most popular piece of art in a museum by observing how often certain floor tiles had to be replaced.

The goal of this approach is the capture of specific measures by mining *indirect* measures based on the accretion and erosion behaviors of populations as they move around their daily lives. *Accretion* behaviors are those that involve the mining of created artifacts. This may involve such things as cataloging what people throw in their trash cans and put out on the curb to understand patterns of food consumption (Rathje and Murphy 2001) and tracking status updates on Twitter to understand the spread of disease (Sadilek et al. 2012). The converse, *erosion* patterns, track physical wear and tear on an object. The replaced floor tiles are an example, as is studying the so-called cow paths— physical traces made by populations as they find the best way to get from one place to another (often not the designed, paved solution). Although the notion of "erosion" is less obvious in digital contexts, systems such as Waze (2014) have similarly analyzed individual paths (as measured by cell-phone traces) to identify the fastest route from place to place. The Edit-wear and Read-wear system proposed by Hill et al. (1992) similarly captured where in a document individuals were spending their time reading or editing.

There are many modern examples for passive crowd work that utilize data from social media. Twitter, Facebook, foursquare, Flickr, and other social media have been used as sources of behavioral traces that are utilized in empirical studies and in the design of systems. A popular application has been the identification of leading indicators for various things, among them the spread of disease (Sadilek et al. 2012) and the outcomes of elections (Livne et al. 2011). As individuals signal their health (e.g., "high fever today, staying home") or their political opinions (e.g., "just voted for Obama,") through social media channels, this information can be used to predict the future value of some variable (e.g., number of infections or who will win the election).

Other systems have demonstrated the ability to generate sophisticated labels for physical places by passively observing the traces of individuals. For example, the Livehoods project (Cranshaw et al. 2012) utilizes foursquare check-ins to build refined models of geographically based communities, which are often different from the labeled neighborhoods on a map. As individuals wander in their daily lives and report their location to foursquare, the project is able to identify patterns of check-ins across a larger population and to identify those new neighborhood structures. Similar projects have utilized geotagged data to identify where tourists go (Flickr 2010) and identify place "semantics" using tagged (both in the textual and geographical sense) images (Rattenbury et al. 2007).

Passive crowd work has also been used as a means for providing technical support. For example, the HelpMeOut system (Hartmann et al. 2010) used instrumented Integrated Development Environments (IDEs) as way of logging a developer's reaction to a problem (e.g., an error. By logging the error and the fix, the system could build a database of recommendations that could be provided to future developers encountering the same error. The Codex system (Fast et al. 2014) identified common programing idioms by analyzing the code produced by developers (millions of lines of Ruby code) to provide labels and warnings to future developers. Query-Feature Graphs (Fourney et al. 2011) mined search logs for common tasks in an end-user application (the image-editing program GIMP). Often people would issue queries such as "How do I remove red-eyes in GIMP?" The system found these common queries, then, by mining the Web, identified common commands that were used in response documents. This allowed the system to automatically suggest commands that high-level end users would be likely to need. Command-Space (Adar et al. 2014) extended this idea by jointly modeling system features (in this case, things that Adobe Photoshop provided—filters, editing tools, menu options, and so on) and natural language. This information was mined from "found" text that included tutorials, message forums, and other traces left by end users on the Web. By recognizing system features in text and utilizing a distributed vector representation, a number of translations could be supported—for example, finding related features, searching for features based on text, search by analogy, and identifying likely uses of a feature.

System design
Although platforms for passive crowd work are attractive in that they don't require intervention or disrupting the user, they must be designed carefully. The *inference gap* reflects the fact that many of the observed behaviors are quite distant from the actual work we would like to see performed. That is, a Twitter user may say "I'm feel terrible today," or a Google searcher may be looking for "fever medication," but what the system requester would really like to know is whether the person is sick with the flu today. The further the "instrument" is from what is being measured, the more difficult it is to make the inference. Additionally, many systems and behaviors change over time (for example, the results found by a search engine change, a social media system is used differently, or the interface adds additional functions or removes others). Consequently, a great deal of care is necessary to ensure that the models and inferences remain predictive (Lazer et al. 2014). Ideally, a passive crowd system would measure behavior in the closest way possible to what is actually the target of measurement and would allow any inference to be updated.

The *reactivity* of the passive solution—that is, when mined behavioral data are used in a feedback loop inside the system—also should be considered. For example, a frequently clicked on search result will move to the top of the search engine's results page. However, this will reduce the chance that other, potentially better pages will be identified. Similarly, if the public is aware that tweets are being used to predict elections, their tweeting behavior may change and the accuracy of forecasting may suffer (Gayo-Avello 2013).

Ethics
The ethical issues associated with passive crowd work are somewhat different than those associated with active work. Those producing work aren't likely to be aware that their traces are being used or for what purposes. The decision as to when and how this information is shared is critical. Facebook, for example, ran a field experiment to determine whether seeing positively or negatively valenced emotion words in friends' status updates would cause users to use similarly

valenced words in their own posts (Kramer et al. 2014). They called this a test of "emotional contagion." To test their hypothesis, they randomly hid some status updates matching a list of positive or negative emotion words from the newsfeed, and found a small effect of friends' using more words of the same kind if the status update was included in the newsfeed. When Facebook published that result, the mass media and several researchers blasted Facebook for running experiments on emotional contagion without informed consent. The goal of learning about people was overshadowed for these individuals by ethical concerns about online experimentation. The case is an especially fraught one because Internet companies such as Google and Facebook run such experiments daily to improve their products.

In addition, depending on how much is explained, a collection process that was once non-reactive may no longer be perceived as such. End users being tracked are now aware of the collection and the potential use of their behavioral traces, and they may act differently than they had before. This also opens up the system to creative attacks by search-engine optimizers who may seek to change the way the system operates. Finally, because the workers aren't aware that they are doing work, they are often unpaid (at least, they often don't receive direct compensation). These considerations must be weighed.

Challenges in Crowdsourcing

Human–computer interaction is helping to shape the future of crowdsourcing through its design of the technology that people will use to engage with crowdsourcing as either requesters or crowd workers. Over the past few years, the field has become aware that the problems on which it chooses to focus may have much of an effect on the benefits we stand to gain through crowdsourcing or on how people choose to work in the future. Kittur et al. (2013) engage in a thorough exploration of these challenges, asking what it would take for us to be comfortable with our own children becoming full-time crowd workers.

Since the earliest days of human computation, its proponents have discussed how the eventual goal is to develop hybrid systems that will engage with both humans intelligence drawn from the crowd and machine intelligence realized through artificial intelligence and machine learning. That vision remains, but systems still utilize it in very basic ways. One of our visions for crowdsourcing in the future is one in which truly intelligent systems are developed more quickly by initially create crowd-powered systems and then using them as scaffolding to gradually move to fully automated approaches.

Crowdsourcing has traditionally worked best, although not exclusively, for problems that required little expertise. A challenge for the future is to widen the scope of problems that it is possible to solve with crowdsourcing by engaging with expert crowds, by embedding needed expertise in the tools non-expert crowds use, or by using a flexible combination of the two (see, e.g., Retelny et al. 2014).

As more people participate as crowd workers, it is becoming increasingly important to understand this component of the labor force and what tools might be helpful to both requesters and workers. Workers on many crowd marketplaces face inefficiencies that could be improved with better tools, such as finding tasks that are well suited to their skills. It is also difficult for workers today to be rewarded over time for acquiring expertise in a particular kind of crowd work.

Conclusion

Human–computer interaction has contributed to crowdsourcing by creating tools that allow different stakeholders to participate more easily or more powerfully, and understanding how people participate in order to shape a brighter future for crowd work. One of the reasons that crowdsourcing is interesting is because technology is allowing groups to work together in ways that weren't feasible a few years ago. Two challenges for the future are to ensure that requesters and workers are able to realize the potential of crowdsourcing without succumbing to its potential downsides and to continue to improve the systems enabling all of this so that even more is possible.

Acknowledgments

We would like to thank Rob Miller for his early brainstorms and for his suggestions on how the chapter should be organized.

References

Ackerman, Mark S., and Thomas W. Malone. 1990. Answer Garden: A tool for growing organizational memory. In *Proceedings of the ACM SIGOIS and IEEE CS TC-OA Conference on Office Information Systems*. ACM.

Adar, Eytan, Mira Dontcheva, and Gierad Laput. 2014. CommandSpace: Modeling the relationships between tasks, descriptions and features. In *Proceedings of the 27th Annual ACM Symposium on User Interface Software and Technology*. ACM.

Ahmad, Salman, Alexis Battle, Zahan Malkani, and Sepander Kamvar. 2011. The jab-berwocky programming environment for structured social computing. In *Proceedings of the 24th Annual ACM Symposium on User Interface Software and Technology*. ACM.

AirSage. 2014. AirSage. http://www.airsage.com.

Antin, Judd, and Aaron Shaw. 2012. Social desirability bias and self-reports of motivation: A cross-cultural study of Amazon Mechanical Turk in the US and India. In *Proceedings of the 2012 ACM Conference on Human Factors in Computing Systems*. ACM.

Beenen, Gerard, Kimberly Ling, Xiaoqing Wang, Klarissa Chang, Dan Frankowski, Paul Resnick, and Robert E. Kraut. 2004. Using Social Psychology to Motivate Contributions to Online Communities. In *Proceedings of the 2004 ACM Conference on Computer Supported Cooperative Work*. ACM.

Bernstein, Michael S., Joel Brandt, Robert C. Miller, and David R. Karger. 2011. Crowds in two seconds: Enabling realtime crowd-powered interfaces. In *Proceedings of the 24th Annual ACM Symposium on User Interface Software and Technology*. ACM.

Bernstein, Michael S., Greg Little, Robert C. Miller, Björn Hartmann, and Mark S. Ackerman, David R. Karger, David Crowell, and Katrina Panovich. 2010a. Soylent: A word processor with a crowd inside. In *Proceedings of the 23nd Annual ACM Symposium on User Interface Software and Technology*. ACM.

Bernstein, Michael S., Desney Tan, Greg Smith, Mary Czerwinski, and Eric Horvitz. 2010b. Personalization via friendsourcing. *ACM Transactions on Computer-Human Interaction* 17 (2): article 6.

Bigham, Jeffrey P., Chandrika Jayant, Hanjie Ji, Greg Little, Andrew Miller, Robert C. Miller, Robin Miller, et al. 2010. VizWiz: Nearly real-time answers to visual questions. In *Proceedings of the 23nd Annual ACM Symposium on User Interface Software and Technology*. ACM.

Burbank, Noah, Debojyoti Dutta, Ashish Goel, David Lee, Eli Marschner, and Narayanan Shivakumar. 2011. Widescope: A social platform for serious conversations on the Web. arXiv preprint arXiv:1111.1958 (2011).

Cheng, Justin, and Michael Bernstein. 2014. Catalyst: Triggering collective action with thresholds. In *Proceedings of the 17th ACM Conference on Computer Supported Cooperative Work & Social Computing*. ACM.

Chi, Ed H. 2009. A position paper on living laboratories: Rethinking ecological designs and experimentation in human–computer interaction. In *Human–Computer Interaction: New Trends*, ed. J. Jacko. Springer.

Chilton, Lydia B., Greg Little, and Darren Edge. Daniel S. Weld, and James A. Landay. 2013. Cascade: Crowdsourcing taxonomy creation. In *Proceedings of the 2013 ACM Annual Conference on Human Factors in Computing Systems*. ACM.

Chilton, Lydia B. Clayton T Sims, Max Goldman, Greg Little, and Robert C Miller. 2009. Seaweed: A Web application for designing economic games. In *Proceedings of the ACM SIGKDD Workshop on Human Computation*. ACM.

Cooper, Seth, Firas Khatib, Adrien Treuille, Janos Barbero, Jeehyung Lee, Michael Beenen, Andrew Leaver-Fay, David Baker, Zoran Popović, and Foldit players. 2010. Predicting protein structures with a multiplayer online game. *Nature* 466 (7307): 756–760.

Cranshaw, Justin, and Aniket Kittur. 2011. The polymath project: Lessons from a success-ful online collaboration in mathematics. In *Proceedings of the SIGCHI Conference on Human Factors in Computing Systems*. ACM.

Cranshaw, Justin, Raz Schwartz, Jason I. Hong, and Norman M. Sadeh. 2012. The Live-hoods Project: Utilizing social media to understand the dynamics of a city. In *Proceedings of the Sixth International AAAI Conference on Weblogs and Social Media*. AAAI Press.

Culliss, G. 2000. Method for organizing information (June 20, 2000). US Patent 6,078,91 (http://www.google.com/patents/US6078916).

Evans, Brynn M., and Ed H. Chi. 2008. Towards a model of understanding social search. In *Proceedings of the 2008 ACM Conference on Computer Supported Cooperative Work*. ACM.

Faridani, Siamak, Ephrat Bitton, Kimiko Ryokai, and Ken Goldberg. 2010. Opinion space: A scalable tool for browsing online comments. In *Proceedings of the SIGCHI Conference on Human Factors in Computing Systems*. ACM.

Fast, Ethan, Daniel Steffee, Lucy Wang, Joel Brandt, and Michael S. Bernstein. 2014. Emergent, crowd-scale programming practice in the IDE. In *Proceedings of the SIGCHI Conference on Human Factors in Computing Systems*. ACM.

Flickr. 2010. Locals and Tourists—a set on Flickr. https://www.flickr.com/photos/walk-ingsf/sets/72157624209158632/.

Foldit. 2008. Foldit: Solve Puzzles for Science. http://fold.it.

Fourney, Adam, Richard Mann, and Michael Terry. 2011. Query-feature graphs: Bridging user vocabulary and system functionality. In *Proceedings of the 24th Annual ACM Sympo-sium on User Interface Software and Technology*. ACM.

Gayo-Avello, Adniel. 2013. A meta-analysis of state-of-the-art electoral prediction from Twitter data. *Social Science Computer Review* 31 (6): 649–679.

Gerber, Elizabeth M., Julie S. Hui, and Pei-Yi Kuo. 2012. Crowdfunding: Why people are motivated to post and fund projects on crowdfunding platforms. In Proceedings of the International Workshop on Design, Influence, and Social Technologies: Techniques, Impacts and Ethics.

Gilbert, Eric. 2013. Widespread underprovision on Reddit. In *Proceedings of the 2013 Conference on Computer Supported Cooperative Work*. ACM.

Hacker, Severin, and Luis von Ahn. 2012. *Duolingo*. www.duolingo.com.

Hartmann, Björn, Daniel MacDougall, Joel Brandt, and Scott R. Klemmer. 2010. What would other programmers do? Suggesting solutions to error messages. In *Proceedings of the SIGCHI Conference on Human Factors in Computing Systems*. ACM.

Heer, Jeffrey, Fernanda B. Viégas, and Martin Wattenberg. 2007. Voyagers and voyeurs: Supporting asynchronous collaborative information visualization. In *Proceedings of the SIGCHI Conference on Human Factors in Computing Systems*. ACM.

Hellerstein, Joseph M., and David L. Tennenhouse. 2011. Searching for Jim Gray: A tech-nical overview. *Communications of the ACM* 54 (7): 77–87.

Hill, Benjamin Mako, and Andrés Monroy-Hernández. 2013. The cost of collaboration for code and art: Evidence from a remixing community. In *Proceedings of the 2013 Confer-ence on Computer Supported Cooperative Work*. ACM.

Hill, William C., James D. Hollan, Dave Wroblewski, and Tim McCandless. 1992. Edit wear and read wear. In *Proceedings of the SIGCHI Conference on Human Factors in Computing Systems*. ACM.

Horowitz, Damon, and Sepandar D. Kamvar. 2010. The anatomy of a large-scale social search engine. In *Proceedings of the 19th international conference on World Wide Web*. ACM.

Irani, Lilly C., and M. Silberman. 2013. Turkopticon: Interrupting worker invisibility in Amazon Mechanical Turk. In *Proceedings of the SIGCHI Conference on Human Factors in Computing Systems*. ACM.

Kanefsky, B., N. G. Barlow, and V. C. Gulick. 2001. Can distributed volunteers accomplish massive data analysis tasks? In *Lunar and Planetary Institute Science Conference Abstracts*, volume 32.

Keegan, Brian, and Darren Gergle. 2010. Egalitarians at the gate: One-sided gatekeeping practices in social media. In *Proceedings of the 2010 ACM Conference on Computer Supported Cooperative Work*. ACM.

Khatib, Firas, Frank DiMaio, Seth Cooper, Maciej Kazmierczyk, Miroslaw Gilski, Szymon Krzywda, Helena Zabranska, et al. 2011. Crystal structure of a monomeric retroviral protease solved by protein folding game players. *Nature Structural & Molecular Biology* 18 (10): 1175–1177.

Kim, Joy, Justin Cheng, and Michael S. Bernstein. 2014. Ensemble: Exploring complementary strengths of leaders and crowds in creative collaboration. In *Proceedings of the 17th ACM Conference on Computer Supported Cooperative Work and Social Computing*. ACM.

Kittur, Aniket, and Robert E. Kraut. 2008. Harnessing the wisdom of crowds in Wikipedia: Quality through coordination. In *Proceedings of the 2008 ACM Conference on Computer Supported Cooperative Work*. ACM.

Kittur, Aniket, Jeffrey V. Nickerson, Michael Bernstein, Elizabeth Gerber, Aaron Shaw, John Zimmerman, Matt Lease, and John Horton. 2013. The future of crowd work. In *Proceedings of the 2013 Conference on Computer Supported Cooperative Work*. ACM.

Kittur, Aniket, Boris Smus, Susheel Khamkar, and Robert E. Kraut. 2011. CrowdForge: Crowdsourcing complex work. In *Proceedings of the 24th Annual ACM Symposium on User Interface Software and Technology*. ACM.

Kittur, Aniket, Bongwon Suh, Bryan A. Pendleton, and Ed H. Chi. 2007. He says, she says: Conflict and coordination in Wikipedia. In *Proceedings of the 2007 SIGCHI Conference on Human Factors in Computing Systems*. ACM.

Kramer, Adam D. I., Jamie E. Guillory, and Jeffrey T. Hancock. 2014. Experimental evidence of massive-scale emotional contagion through social networks. *Proceedings of the National Academy of Sciences* 111 (24): 8788–8790.

Kraut, Robert E., and Paul Resnick. 2012. *Building Successful Online Communities: Evidence-Based Social Design*. MIT Press.

Kriplean, Travis, Jonathan T. Morgan, Deen Freelon, Alan Borning, and Lance Bennett. 2011a. ConsiderIt: Improving structured public deliberation. In *CHI '11 Extended Abstracts on Human Factors in Computing Systems*. ACM.

Kriplean, T., M. Toomim, J. T. Morgan, A. Borning, and A. J. Ko. 2011b. REFLECT: Supporting active listening and grounding on the Web through restatement. In Proceedings of the Conference on Computer Supported Cooperative Work, Hangzhou.

Kulkarni, Anand P., Matthew Can, and Björn Hartmann. 2012. Turkomatic: Automatic recursive task and workflow design for mechanical Turk. In *Proceedings of the ACM 2012 conference on Computer Supported Cooperative Work*. ACM.

Lasecki, Walter S., Kyle I. Murray, and Samuel White. Robert C Miller, and Jeffrey P. Bigham. 2011. Real-time crowd control of existing interfaces. In *Proceedings of the 24th Annual ACM Symposium on User Interface Software and Technology*. ACM.

Lasecki, Walter S., Christopher Miller, Adam Sadilek, Andrew Abumoussa, Donato Borrello, Raja Kushalnagar, and Jeffrey Bigham. 2012. Real-time captioning by groups of non-experts. In *Proceedings of the 25th Annual ACM Symposium on User Interface Software and Technology*. ACM.

Lasecki, Walter S., Christopher D. Miller, and Jeffrey P. Bigham. 2013. Warping time for more effective real-time crowdsourcing. In *Proceedings of the 2013 ACM Annual Conference on Human Factors in Computing Systems*. ACM.

Lave, Jean, and Etienne Wenger. 1991. *Situated Learning: Legitimate Peripheral Participation*. Cambridge University Press.

Lazer, David M., Ryan Kennedy, Gary King, and Alessandro Vespignani. 2014. The parable of Google Flu: Traps in Big Data analysis. *Science* 343 (6176): 1203–1205.

Lease, Matthew, Jessica Hullman, Jeffrey P. Bigham, Michael Bernstein, J. Kim, W. Lasecki, S. Bakhshi, T. Mitra, and R. C. Miller. 2013. Mechanical Turk is not anonymous. Social Science Research Network.

Lintott, Chris J., Kevin Schawinski, Anže Slosar, Kate Land, Steven Bamford, Daniel Thomas, M. Jordan Raddick, Robert C. Nichol, Alex Szalay, Dan Andreescu, Phil Murray, and Jan Vandenberg. 2008. Galaxy Zoo: Morphologies derived from visual inspection of galaxies from the Sloan Digital Sky Survey. *Monthly Notices of the Royal Astronomical Society* 389 (3): 1179–1189.

Little, Greg, Lydia B. Chilton, Max Goldman, and Robert C. Miller. 2009. TurKit: Tools for iterative tasks on mechanical Turk. In *Proceedings of the ACM SIGKDD Workshop on Human Computation*. ACM.

Livne, Avishay, Matthew P. Simmons, Eytan Adar, and Lada A. Adamic. 2011. The party is over here: Structure and content in the 2010 election. In Proceedings of the Fifth International AAAI Conference on Weblogs and Social Media.

Louv, Richard, John W. Fitzpatrick, Janis L. Dickinson, and Rick Bonney. 2012. *Citizen Science: Public Participation in Environmental Research*. Cornell University Press.

Luther, Kurt, and Amy Bruckman. 2008. Leadership in online creative collaboration. In *Proceedings of the 2008 ACM Conference on Computer Supported Cooperative Work*. ACM.

Luther, Kurt, Kelly Caine, Kevin Ziegler, and Amy Bruckman. 2010. Why it works (when it works): Success factors in online creative collaboration. In *Proceedings of the 16th ACM International Conference on Supporting Group Work*. ACM.

Luther, Kurt, Casey Fiesler, and Amy Bruckman. 2013. Redistributing leadership in online creative collaboration. In *Proceedings of the 2013 Conference on Computer Supported Cooperative Work*. ACM.

Mason, Winter, and Duncan J. Watts. 2010. Financial incentives and the performance of crowds. *ACM SigKDD Explorations Newsletter* 11 (2): 100–108.

Morgan, Jonathan T., Siko Bouterse, Heather Walls, and Sarah Stierch. 2013. Tea and sympathy: Crafting positive new user experiences on Wikipedia. In *Proceedings of the 2013 Conference on Computer Supported Cooperative Work*. ACM.

Morris, Meredith Ringel, Jaime Teevan, and Katrina Panovich. 2010. What do people ask their social networks, and why? A survey study of status message Q&A behavior. In *Proceedings of the SIGCHI Conference on Human Factors in Computing Systems*. ACM.

Munson, Sean A., Stephanie Y. Lee, and Paul Resnick. 2013. Encouraging reading of diverse political viewpoints with a browser widget. In Proceedings of the Seventh International AAAI Conference on Weblogs and Social Media.

Naim, Iftehar, Daniel Gildea, Walter Lasecki, and Jeffrey P. Bigham. 2013. Text alignment for real-time crowd captioning. In *Proceedings of NAACL-HLT*.

Nalimov, E. V., C. Wirth, G. M. C. Haworth, et al. 1999. KQQKQQ and the Kasparov-World Game. *ICGA Journal* 22 (4): 195–212.

Noronha, Jon, Eric Hysen, Haoqi Zhang, and Krzysztof Z. Gajos. 2011. Platemate: Crowdsourcing nutritional analysis from food photographs. In *Proceedings of the 24th Annual ACM Symposium on User Interface Software and Technology*. ACM.

Okolloh, Ory. 2009. Ushahidi, or "testimony": Web 2.0 tools for crowdsourcing crisis information. *Participatory Learning and Action* 59 (1): 65–70.

Panciera, Katherine, Aaron Halfaker, and Loren Terveen. 2009. Wikipedians are born, not made: A study of power editors on Wikipedia. In *Proceedings of the ACM 2009 International Conference on Supporting Group Work*. ACM.

Preece, Jennifer, and Ben Shneiderman. 2009. The reader-to-leader framework: Motivating technology-mediated social participation. *AIS Transactions on Human–Computer Interaction* 1 (1): 13–32.

Quinn, Alexander J., and Benjamin B. Bederson. 2014. AskSheet: Efficient human computation for decision making with spreadsheets. In *Proceedings of the 17th ACM Conference on Computer Supported Cooperative Work and Social Computing*. ACM.

Rathje, William L., and Cullen Murphy. 2001. *Rubbish! The Archaeology of Garbage*. University of Arizona Press.

Rattenbury, Tye, Nathaniel Good, and Mor Naaman. 2007. Towards automatic extraction of event and place semantics from flickr tags. In *Proceedings of the 30th Annual International ACM SIGIR Conference on Research and Development in Information Retrieval*. ACM.

Resnick, Mitchel, John Maloney, Andrés Monroy-Hernández, Natalie Rusk, Evelyn Eastmond, Karen Brennan, Amon Millner, et al. 2009. Scratch: Programming for all. *Communications of the ACM* 52 (11): 60–67.

Retelny, Daniela, Sebastien Robaszkiewicz, Alexandra To, Walter Lasecki, Jay Patel, Negar Rahmati, Tulsee Doshi, Melissa Valentine, and Michael S. Bernstein. 2014. Expert crowdsourcing with flash teams. In Proceedings of the 27th annual ACM Symposium on User Interface Software and Technology. ACM.

Russell, Brian C., Antonio Torralba, Kevin P. Murphy, and William T. Freeman. 2008. LabelMe: A database and Web-based tool for image annotation. *International Journal of Computer Vision* 77 (1): 157–173.

Ryan, Richard M., and Edward L. Deci. 2000. Self-determination theory and the facilitation of intrinsic motivation, social development, and well-being. *American Psychologist* 55 (1), 68–78.

Rzeszotarski, Jeffrey, and Aniket Kittur. 2012. CrowdScape: Interactively visualizing user behavior and output. In *Proceedings of the 25th Annual ACM Symposium on User Interface Software and Technology*. ACM.

Sadilek, Adam, Henry A. Kautz, and Vincent Silenzio. 2012. Modeling spread of disease from social interactions. In Proceedings of the Sixth International AAAI Conference on Weblogs and Social Media.

Salehi, Niloufar, Lilly Irani, Michael Bernstein, Ali Alkhatib, Eva Ogbe, Kristy Milliland, and Clickhappier. 2015. We are Dynamo: Overcoming stalling and friction in collective action for crowd workers. In *Proceedings of the SIGCHI Conference on Human Factors in Computing Systems*.

Salganik, Matthew J., and Duncan J. Watts. 2009. Web-based experiments for the study of collective social dynamics in cultural markets. *Topics in Cognitive Science* 1 (3): 439–468.

Settles, Burr, and Steven Dow. 2013. Let's get together: The formation and success of online creative collaborations. In *Proceedings of the 2013 ACM Annual Conference on Human Factors in Computing Systems*. ACM.

Singh, Push, Thomas Lin, Erik T. Mueller, Grace Lim, Travell Perkins, and Li Zhu Wan. 2002. Open mind common sense: Knowledge acquisition from the general public. In *On the Move to Meaningful Internet Systems 2002: CoopIS, DOA, and ODBASE*, ed. Z. Tari. Springer.

Starbird, Kate. 2013. Delivering patients to Sacré Coeur: Collective intelligence in digital volunteer communities. In *Proceedings of the SIGCHI Conference on Human Factors in Computing Systems*. ACM.

Tuite, Kathleen, Noah Snavely, Dun-yu Hsiao, Nadine Tabing, and Zoran Popović. 2011. PhotoCity: Training experts at large-scale image acquisition through a competitive game. In *Proceedings of the SIGCHI Conference on Human Factors in Computing Systems*. ACM.

Viégas, Fernanda B., Martin Wattenberg, Frank Van Ham, Jesse Kriss, and Matt McKeon. 2007. Many eyes: A site for visualization at internet scale. *Visualization and Computer Graphics. IEEE Transactions on* 13 (6): 1121–1128.

Vieweg, Sarah. Amanda L. Hughes, Kate Starbird, and Leysia Palen. 2010. Microblogging during two natural hazards events: What Twitter may contribute to situational awareness. In *Proceedings of the SIGCHI Conference on Human Factors in Computing Systems*. ACM.

von Ahn, Luis, and Laura Dabbish. 2004. Labeling images with a computer game. In *Proceedings of the SIGCHI Conference on Human Factors in Computing Systems*. ACM, 319–326.

von Ahn, Luis, Benjamin Maurer, Colin McMillen, David Abraham, and Manuel Blum. 2008. reCAPTCHA: Human-Based Character Recognition via Web Security Measures. *Science* 321 (5895): 1465–1468.

Waze. 2014. Waze. http://waze.com.

Webb, E. J., D. T. Campbell, R. D. Schwartz, and L. Sechrest. 1999. *Unobtrusive Measures*. SAGE.

Artificial Intelligence

Editors' Introduction

The field of artificial intelligence (AI) strives to create computers that can reason on their own. The field formed with a goal to try and match human intelligence using a computer, and in the intervening years it has evolved into a highly successful example of data-driven engineering. Artificial intelligence algorithms typically capture raw data about the world, then use those data to inductively build models of previous experiences and react intelligently the next time they encounter similar situations. Today artificial intelligence powers email spam filters, search engines, voice interfaces, and news-feed rankings for social networks such as Facebook.

The discipline has often rallied collective intelligence around goals that remain beyond algorithmic reach. NASA's crowdsourced Clickworkers, for example, gathered to identify craters on Mars because computer vision software could not do so reliably. Likewise, Amazon CEO Jeff Bezos originally billed the company's Mechanical Turk as "artificial artificial intelligence." Sometimes artificial intelligence even competes indirectly against collective intelligence, as when the chess grandmaster Garry Kasparov separately challenged the world's other human players (a challenge he won) and IBM's Deep Blue algorithm (a challenge he lost).

Artificial intelligence and collective intelligence need not be in tension—together, humans and computers could be more powerful than either humans or computers alone. Could artificial intelligence play an active role in collective intelligence, rather than seeking a *deus ex crowd* when the task is too difficult? As Dan Weld, Mausam, Christopher Lin, and Jonathan Bragg discuss in the chapter that follows, data-driven and model-driven artificial intelligence techniques promise to make collectives even more intelligent. Using directed crowdsourcing such as on Amazon Mechanical Turk as an example, Weld and colleagues provide a second perspective on the material that Bigham and colleagues introduced in the chapter on human–computer interaction.

The most common modern method in artificial intelligence, and thus the method covered most thoroughly in the chapter, is statistical modeling. Typically an AI system requires the creation of some sort of model of the world. That model must be broken down into *features* such as word counts, observed low-level pixel patterns, or human judgments. The model also requires an *objective*—for example, classifying email messages as spam or not, or minimizing the human cost of labeling a data point correctly. The system's goal is to maximize (or minimize) the objective: correctly classify as many spam messages as possible, or find the lowest-cost way to label the data. To solve this problem, the system learns from experience. A developer will provide many hundreds or thousands of example inputs' features and, depending on the goal, their ground-truth answers. Algorithms build a model of the world from the low-level features in these examples, then apply that model to new inputs.

More broadly, foundational work in artificial intelligence spans many years and many models. Popular domains include computer vision (image understanding), natural language processing (reading unstructured text), and robotics. In its early foundations, the field considered human-brain-inspired models such as the neural network. Another major historical movement was *expert systems*, where a human knowledge engineer attempted to draw out and then codify a set of rules that a domain expert would apply in a domain such as medicine. These rule-based systems were explicit; eventually, this model gave way to today's data-driven approaches that sacrifice easy interpretation for higher accuracy.

In the chapter that follows, Weld and colleagues focus specifically on how learning from experience can help collective intelligence use only as many people as it needs to. One side of this coin is estimating confidence in contributors' answers: Do we need another answer to feel confident? Do those participants' previous answers tell us anything about how much we can trust them? The other side of the coin integrates AI methods into collective intelligence systems: Could an artificial intelligence agent make guesses as to certain classifications, essentially becoming part of the collective itself?

Recommended Readings

Stuart Russell and Peter Norvig. 1995. *Artificial Intelligence: A Modern Approach.* Prentice-Hall.

This broad introduction to the field and its techniques covers topics ranging from planning (e.g., for chess-playing computers or robots) to neural networks and first-order logic.

A. P. Dawid, and A. M. Skene. 1979. Maximum likelihood estimation of observer error-rates using the EM algorithm. *Journal of Applied Statistics* 28 (1): 20–28.

This paper in applied statistics has become the backbone model for the optimal use of crowdsourcing. It asks: Given a set of forced-choice guesses (here from patient records, but equivalently from a noisy crowd), how certain am I that I know the correct answer to the question? This algorithm looks at which workers are agreeing with the majority in order to understand which workers' answers are likely to be of high quality. It weights their votes higher, then recalculates what it thinks the right answer is using those weights. That process then loops, weighting workers by estimated correct answers, then weighting answers on the basis of estimated worker quality until the entire process converges. This method has been used by many companies and researchers to understand how many of their workers must answer each question.

Peng Dai, Mausam, Christopher Lin, and Dan S. Weld. 2013. POMDP-based control of workflows for crowdsourcing. *Artificial Intelligence* 202: 52–85.

TurKontrol extended the approach of Get Another Label from single questions to entire workflows. Should we have another worker iterate on this answer, or should we take a vote on which previous iteration to keep? How many times should we loop this process? TurKontrol explicitly models the overall cost of the workflow, including any decisions to branch or loop, and attempts to minimize it at runtime in view of the answers obtained so far. This paper marked one of the first attempts to integrate statistical machine learning with crowdsourcing workflows.

Ece Kamar, Severin Hacker, and Eric Horvitz. 2012. Combining human and machine intelligence in large-scale crowdsourcing. In *Proceedings of the 11th International Conference on Autonomous Agents and Multiagent Systems.* International Foundation for Autonomous Agents and Multiagent Systems.

Artificial intelligence can become part of the collective. There is little point asking people to label data when AI techniques have high confidence that they answered the question correctly. Since AI is much more scalable than human effort, this paper demonstrates how to integrate AI algorithms with human judgments. Doing so allows people to focus on the toughest questions, and exposes the potential of complementarity between the crowd and the algorithm.

Rion Snow, Brendan O'Connor, Daniel Jurafsky, and Andrew Y. Ng. 2008. Cheap and fast—but is it good? Evaluating non-expert annotations for natural language tasks. In *Proceedings of the Conference on Empirical Methods in Natural Language Processing.* Association for Computational Linguistics.

Artificial intelligence research requires large amounts of data. Snow et al. were among the first researchers to ask whether paid crowdsourcing platforms might serve as a readily available source of such data. They compared annotations from Amazon Mechanical Turk with expert annotations and determined that, for many goals, paid microtask workers were up to the task. Their paper has been cited many times to support claims that Amazon Mechanical Turk can be a reliable source of training data for algorithms.

Artificial Intelligence and Collective Intelligence

Daniel S. Weld, Mausam, Christopher H. Lin, and Jonathan Bragg

The vision of artificial intelligence (AI) is often manifested through an autonomous software module (agent) in a complex and uncertain environment. The agent is capable of thinking ahead and acting for long periods of time in accordance with its goals/objectives. It is also capable of learning and refining its understanding of the world. The agent may accomplish this on the basis of its own experience or from feedback provided by humans. Famous recent examples include self-driving cars (Thrun 2006) and Watson, the computer program that IBM developed to compete with humans at *Jeopardy* (Ferrucci et al. 2010). In this chapter we explore the immense value of AI techniques for collective intelligence, including ways to make interactions between large numbers of humans more efficient.

If one begins by defining collective intelligence as "groups of individuals acting collectively in an intelligent manner," one soon wishes to nail down the meaning of *individual*. In this chapter, individuals may be software agents and/or people and the collective may consist of a mixture of both. The rise of collective intelligence allows novel possibilities of seamlessly integrating machine and human intelligence at a large scale—one of the "holy grails" of AI (known in the literature as *mixed-initiative systems*) (Horvitz 2007)). In this chapter we focus on one such integration: the use of machine intelligence for the management of *crowdsourcing* platforms (Weld, Mausam, and Dai 2011).

In crowdsourcing (a special case of collective intelligence), a third party (called the *requester*) with some internal objective solicits a group of individuals (called *workers*) to perform a set of interrelated tasks in service of that objective. The requester's objective may be expressed in the form of a utility function to be maximized. For example, a requester might wish to obtain labels for a large set of images; in this case, the requester's utility function might be the average quality of labels subject to a constraint that no more than X dollars be spent on paying workers. We assume that the workers act independently, interacting only through the shared tasks. Each worker has an individual utility function, which is often different from the collective's utility function. Furthermore, we assume that the workers' utility functions are independent of one another. The AI subfield of *multi-agent systems* considers even richer models, in which individual agents may reason about the objectives of other agents, negotiate, and bargain with one another (Weiss 2013).

There are two natural points of connection between AI and crowdsourcing: AI for crowdsourcing and crowdsourcing for AI. Although this chapter centers on the former, we note that in recent years crowdsourcing has had a significant impact on AI research as well—a great many projects use crowdsourcing to label training sets as input for data-hungry supervised learning algorithms (Snow et al. 2008; Callison-Burch 2009; Hsueh, Melville, and Sindhwani 2009).

Why does crowdsourcing need AI? Crowdsourcing is an effective medium for assembling (usually virtually) a large number of workers who assist with a common goal. This allows for creative new applications that use the wisdom of crowds or the round-the-clock availability of people (see, e.g., Bigham et al. 2010). At the same time, the sheer volume of tasks and the widely varying skills and abilities of workers typically make it infeasible to manually manage task allocation and quality control. Moreover, designing crowdsourced interfaces and workflows to accomplish a new task remains difficult and expensive. For example, often a task may get routed to a worker who isn't interested in it, or to one who lacks skills that the task requires. Different tasks may require slightly different workflows to achieve high quality. Different task instances may be individually easier or more difficult, requiring less or more work (iterations) on them. These and other challenges necessitate the use of automated techniques for the design and management of crowdsourcing processes.

A long-term vision of AI for crowdsourcing is to enable optimal design of workflows and management of task instances, thereby making crowdsourcing platforms highly efficient, saving thousands of man-hours and millions of dollars, and also making crowdsourcing really easy to use for a novice requester. AI is a natural fit for this vision because, in general, AI algorithms are very good at building models, drawing inferences, and detecting outliers from the data. They are also effective in taking decisions in uncertain environments so as to maximize an objective. In this chapter, we discuss several uses of AI in this space. We describe learning algorithms that model the accuracy of crowd members, aggregation methods for predicting true answers from error-prone and disagreeing workers, and AI control algorithms that choose which tasks to request and which individuals should work on them.

Preliminaries

Because different individuals in a crowdsourced collective may have differing priorities, it's important to be clear on terminology; for that reason, we have provided a brief glossary at the end of the chapter. We'll consider a variety of objectives in the chapter, but the most common objective is to accurately label a set of examples; here the requester needs to choose how many workers should be given a labeling question and how their responses should be aggregated to estimate the best (i.e., the most likely) answer. Usually the requester's objective includes minimizing the number of questions given to workers, either because the workers are paid per question or just to avoid burdening volunteer workers. Sometimes, however, a requester is interested in minimizing *latency* (i.e., in being able to compute an answer quickly); as we will explain in the chapter's final section, that may require additional tasks.

Either way, we focus on algorithms for helping the requester decide what to do. As you'll see, these methods include various types of machine learning, expectation maximization, optimization, policy construction for partially observable Markov decision processes (POMDPs), and reinforcement learning.

The choice of algorithm depends not just on the requester's objective but also on the *labor market*, an economic term we'll use even if the workers are volunteers and not being paid. In some markets, such as Amazon Mechanical Turk, a requester can post tasks and workers get to choose which task they wish to attempt. In other markets, one can directly assign specific tasks to individual workers. In all cases it turns out to be useful to track workers' responses in order to construct a model of their accuracy and maybe which kinds of tasks they enjoy. In theory, a human worker is capable of performing an arbitrarily complex task, but we focus on simple jobs known as *microtasks*—for example, answering multiple-choice questions, writing or editing a short text description, or drawing a bounding box on an image. Often a complex objective can be achieved by a *workflow* made up of these simple tasks. Finally, as we mentioned in the introduction to the chapter, we assume that individual workers are largely independent of one another; as a result, our approaches may not work well if malevolent workers collude in order to deceive the requester. Fortunately, such behavior is extremely rare in practice.

Collective Assessment and Prediction

In crowdsourcing, to account for variation in workers' skills, requesters often ask multiple workers to perform the same question (or related questions) and then aggregate responses to infer the correct answers. In practice, the effectiveness of crowdsourcing can be highly dependent on the method for aggregating responses, and numerous strategies have been investigated. In this section, we describe a set of *post hoc* response aggregation methods that attempt to maximize the amount of information gleaned after question responses have been received.

Suppose that an AI system is given a set of multiple-choice questions, a set of workers, and a set of their responses such that some questions are answered by more than one worker. We assume that the questions are objective (i.e., each has a unique correct answer) and that workers may or may not answer correctly. We also assume that it is more likely that a majority of workers will be correct than that a majority will be incorrect. (If the majority of workers are likely to agree on an incorrect answer, then more sophisticated methods, such as Bayesian Truth Serum, should be used to reveal the right answer. See Prelec and Seung 2007.) AI systems can work without information beyond the workers' proposed answers. The system's objective is to infer the correct answers from the noisy workers' responses.

Although the negative effects of imperfect workers can be mitigated in many ways, the techniques we consider in this section revolve around two common methods. The first is to exploit redundancy, comparing different workers' responses to the same question. Indeed, Snow et al. (2008) found that simple majority voting allowed a crowd of novices to outperform an expert on natural language labeling tasks such as sentiment analysis and judging word similarity. The second approach is to learn and track the skills of the workers. Rather than a simple majority vote, these approaches weigh workers' responses by using models of workers' abilities. In the simplest case such a model might be a single "accuracy" number, but models can grow arbitrarily complex. For example, if one knew that a worker excelled at translating French to English, one might suspect that her English-to-French translations would also be excellent.

Simple Approaches to Collective Assessment

Let us first consider algorithms that improve on simple majority voting by modeling workers' skills. The simplest approach, and one commonly used in practice, uses supervised learning, which gives workers questions for which "gold-standard" answers are already known (Dai, Mausam, and Weld 2011). Workers who fail to correctly answer "gold standard" questions are dismissed or have their weights lowered. To avoid gaming behavior (in which, for example, a worker might answer the first few questions and then, after convincing the system of his or her aptitude, unleash a simple bot to answer the remaining questions, probably with greatly reduced accuracy), it is common to intermix questions with known and unknown answers. However, even this strategy is foiled by scammers building bots that utilize databases of known questions, leading to elaborate strategies for programmatically generating an unbounded number of "gold-standard" answers (Oleson et al. 2011).

Collective Assessment using Expectation Maximization

More sophisticated approaches eschew "gold-standard" questions entirely, instead using unsupervised learning to jointly estimate worker accuracy and consensus answers together. As a first example, consider the work reported in an early paper by Dawid and Skene (1979). Although they originally pose their approach in terms of medical diagnosis, it clearly fits the crowdsourcing model presented above. There is a single question with an unknown correct answer, and there is a parameter $P_w(r \mid a)$, for each worker and each possible response, describing the probability that worker w will give response r when the true answer is a. These probabilities can be seen as a very simple model of workers' abilities—an expert worker would have $P_w(r \mid a)$ close to zero for all $r \neq a$. Dawid and Skene make an important simplification: they assume that the worker's responses are conditionally independent of one another when the true answer is given. In other words, if we already knew the true answer, our estimate of $P_w(r \mid a)$ should not be affected regardless of how other workers answer the question. David and Skene use an iterative expectation-maximization algorithm to estimate which answers are correct at the same time that the algorithm learns the model of workers' accuracies.

Expectation maximization embodies the intuition that a good worker is one whose answers agree with those of other workers (more precisely, that an excellent worker's answers agree with those of other good workers). Unfortunately, this idea presents a chicken-and-egg dilemma: How can you score one worker without already knowing the quality of her peers? Expectation maximization solves this problem by computing better and better estimates until reaching a fixed point. It starts by taking a majority vote and using that to determine an initial guess of the correct answer for each question. Expectation maximization then scores each worker (P_w) on the basis of how many answers she got right. In subsequent iterations, expectation maximization weights each worker's votes on the basis of her score, so better workers (with higher P_w scores) count more. Since weighted votes are likely to produce a different set of correct answers, the next step is to recompute each worker's score. This process repeats until quiescence. As expectation maximization assigns higher weights to good workers and lower weights to poor workers, it allows a single strong worker to overrule multiple weak workers, and thus the predicted answer may no longer be the answer for which a majority voted.

More precisely, expectation maximization is a general method for learning maximum-likelihood estimates of hidden parameters. As figure 5.1 shows, it initializes the probability parameters to random values and then, using the workers' responses, repeats the following steps to convergence:

Expectation Given estimates of all the probabilities, $P_w(r \mid a)$, compute the probability of each possible answer using Bayes' rule and the assumption of conditional independence.
Maximization Given the posterior probability of each possible answer, compute new parameters $P_w(r \mid a)$ that maximize the likelihood of each worker's response.

The model of Dawid and Skene is a relatively simple one, and researchers have created new models to address its various weaknesses. Whitehill et al. (2009) note that workers' responses are not really independent unless they are conditioned on both the correct answer *and* the question's difficulty. To see this, suppose that, on average, students have an 80 percent chance of correctly answering textbook questions. Then we would expect that Jane, in particular, would have an 80 percent chance when confronted with question 13. However, if we were told that all 25 of the other students in the class had gotten the problem

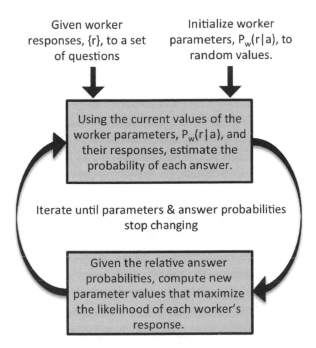

Given worker responses, {r}, to a set of questions

Initialize worker parameters, $P_w(r|a)$, to random values.

Using the current values of the worker parameters, $P_w(r|a)$, and their responses, estimate the probability of each answer.

Iterate until parameters & answer probabilities stop changing

Given the relative answer probabilities, compute new parameter values that maximize the likelihood of each worker's response.

Figure 5.1
Expectation maximization repeats two steps until convergence, alternately estimating the best answers and then updating its model of workers.

wrong, we probably would suspect that question 13 is especially hard and would want to revise our estimate of Jane's chances downward. The model of Dawid and Skene cannot make this inference, but that of Whitehill et al. uses information about workers' errors to update its belief about the difficulty of problems and hence about the accuracy of other workers. The algorithm still follows the expectation-maximization pattern illustrated in figure 5.1, but the probability computations are a bit more complex.

Welinder et al. (2010) take Whitehill's approach a step further, designing a model with general multidimensional parameters. Questions have many features, one of which could be difficulty, and workers are modeled as linear classifiers who make their responses by weighting those features. This allows the model of Welinder et al. to account not only for the skill of the workers and the difficulty of the questions, but also for arbitrary features of the workers and the questions. For instance, they can learn that one worker is particularly good at discerning different types of birds, but only when viewed from the back.

Surprisingly, these features of questions need not be specified in the model *a priori*; the algorithm of Welinder et al. learns the features. Although this leads to excellent performance at answer assessment, it does have a drawback: It may be difficult or impossible for a human to understand the learned model.

All the models we have covered thus far assume that one knows the set of possible answers before giving questions to workers. Lin, Mausam, and Weld (2012a) address cases in which either requesters cannot enumerate all possible answers for the worker or the solution space is infinitely large. Lin et al. use the Chinese Restaurant Process (Aldous 1985), a model that often matches distributions seen in nature. Specifically, they specify a generative probabilistic model in which workers answer a question by returning a previously seen response with probability proportional to the number of other workers who have given that response and returning a novel response with some small fixed probability. The benefits of this model include an ability to handle open-ended questions and an ability to deal with common mistakes that are repeated by multiple workers.

The AI literature abounds with various approaches to *post hoc* response aggregation. Kajino, Tsuboi, and Kashima (2012) note that using expectation maximization to learn parameters can lead to local optima. They propose instead to model the repeated labeling problem as convex optimization, so that a globally optimal solution can always be obtained. Prelec and Seung (2007) develop an algorithm that can find correct answers missed by the majority by asking workers to predict co-workers' mistakes. Liu, Peng, and Ihler (2012) apply belief propagation and mean field approximation, techniques beyond the scope of this book, to perform the inference required to learn correct answers.

Researchers are now beginning to apply the full force of state-of-the-art machine learning algorithms to this problem. To make comparison easier, Sheshadri and Lease (2013) have developed an open-source framework that allows benchmarking of response aggregation methods.

Gradually Moving to Fully Automated Approaches

The next set of approaches seek to do more than simply reconcile multiple answers to a set of questions—they use machine learning to create an autonomous system that can answer questions itself. Raykar et al. (2010) propose a model that not only can learn about workers' abilities and infer correct answers but also can jointly learn a logistic regression classifier that predicts future crowd responses or the answer, thereby eliminating the need to consult human workers in the future. They also use an expectation-maximization pattern to learn their model.

Wauthier and Jordan (2011) relax the idea that a question must have a "correct answer." Instead, they build a model that describes each worker's idiosyncrasies, then use it to predict each worker's response to a future question, q, as a function of features of that question. Since this model handles subjective questions, it is quite powerful. Furthermore, it can be used to answer objective questions by adding an imaginary, virtual worker to specify "gold-standard" answers for some of the questions. Now one can simply use the method of Wauthier et al. to predict how this imaginary always-correct worker would answer future questions. However, that method has some drawbacks. In contrast to the preceding techniques, all of which were unsupervised, it relies on supervised machine learning, which means that it requires that one already know the answers to some of the questions in order to predict the answers to others.

The ultimate objective of Raykar et al. and Wauthier et al. is to replace human workers rather than to derive consensus from their answers. Their methods assume that there exist features for the question that would allow a learned classifier to accurately predict the answer. However, crowdsourcing is often used precisely to answer questions that are beyond AI's state of the art.

Instead of bootstrapping the learning process by using multiple workers to redundantly answer each question, Dekel and Shamir (2009a,b) devise algorithms to limit the influence of bad workers. Specifically, they use workers' responses to train a support vector machine classifier, and they add constraints to the loss function such that no one worker, and no bad worker, can overly influence the learned weights. They also introduce a second technique to completely disregard bad workers, re-solving the questions with the remaining high-quality individuals.

Workflow Optimization

We now shift our focus to optimizing a crowdsourcing process. Specifically, we consider the problem of dynamically controlling the execution of an interrelated set of tasks, which we call a *workflow*.

A requester typically has several objectives that must be jointly optimized. These often include the quality of the output, the total cost of the process (in the case of economically motivated crowdsourcing), and/or other measures of efficiency (number of workers required, total completion time, and so on). Inevitably there are tradeoffs between objectives. For example, one can usually increase the quality of output by enlisting more workers, but that increases cost. On the other hand, one can reduce cost by paying workers less, but that increases latency (Mason and Watts 2009). In this section we describe various approaches useful for crowdsourcing optimization. Although we focus on economically motivated micro-crowdsourcing platforms, such as Amazon Mechanical Turk, the techniques also apply to other platforms, such as Zooniverse (Lintott et al. 2008), where the requester typically wishes to get the highest-quality output from a limited number of volunteers.

Researchers have taken two broad approaches to such optimization. The first is to carefully design an efficient workflow for a given task such that the overall output is of high quality and, at the same time, does not cost much. An example is in sorting of items, such as images. One could use comparison between items, or one could ask workers to rate an item on a numerical scale (Marcus et al. 2011). The latter approach costs less but may not be as accurate. A hybrid scheme that first rates each item independently and later uses a comparison operator on an intelligently chosen subset of items to do fine-grained sorting produces a better tradeoff between cost and quality. There are several other examples of alternative workflows for a task, and they achieve different tradeoffs between cost and quality. These include computing the max from a set of items (Venetis et al. 2012), multiple-choice tasks (Sun and Dance 2012), soliciting translations (Sun, Roy, and Little 2011), writing image descriptions (Little et al. 2010), and taxonomy generation (Chilton et al. 2013). Designing an efficient workflow is usually task dependent. It requires involvement of a domain expert, and typically it entails iterative design, much trial and error, and painstaking focus on instructions. To our knowledge, there are few general principles for this kind of work. The onus is on a human to creatively come up with a good workflow for the task.

A second (and complementary) approach to optimizing crowdsourcing is more computational. It assumes a given workflow (e.g., one developed using the first approach) that has some parameters or decision points. It then uses AI to optimize the workflow by learning the best values of the parameters and controlling the workflow to route a task through various decision points. Since these methods can get somewhat complex, we will begin with the simplest possible example.

Control of a Binary-Vote Workflow

One of the simplest forms of control problem arises in crowdsourcing of a single binary-choice question, where workers provide either a Yes or a No response. Because workers' responses are noisy, a common solution for quality control is to ask multiple workers and aggregate the responses using majority vote or the expectation-maximization approaches described above. But how many workers should be asked for each question? Choosing the optimal number requires making a tradeoff between cost and desired quality.

Typically, a requester either chooses the optimal number on the basis of the available budget or does some initial performance analysis to understand the average ability of the workers and then picks a number to achieve a desired accuracy. These approaches miss an extremely important insight: Not all questions (nor all workers) are equal. A fixed policy sacrifices the ability to shift effort away from easy problems to improve accuracy on hard questions. A superior solution is to perform *dynamic* control—that is, to decide, on the basis of the exact history of work done so far, whether to take another judgment for each question (see figure 5.2).

As a case study, we will discuss work in which Dai, Mausam, and Weld (2010, 2013) modeled the problem of deciding whether to ask for another vote as a partially observable Markov decision process (a popular technique for decision-theoretic optimization). Describing the POMDP representation and solution algorithms in detail is beyond the scope of this chapter (see Kaelbling, Littman, and Cassandra 1998; Poupart 2011), but at a high level a POMDP is a sequential decision-making problem in which all the actions (in our case, whether to ask for another worker's judgment or simply submit the answer) are known, and the set of *possible* states is known, but the exact state (for example, the true answer) is not observable. The POMDP model defines system dynamics in terms of probability distributions for state transitions and observations. It also defines a reward that the agent tries to maximize.

The first step in defining the POMDP is specifying the state space as a pair (a, d), where a denotes the correct answer for the question (Yes or No) and d is the question's difficulty (a number between 0 and 1). Since neither a nor d changes over time, the POMDP transition probability is simply the identity function, making it a relatively simple problem to solve. However, neither the value of a nor that of d can be observed directly; at best, one can maintain a probability distribution (called a *belief*) over the possible values of (a, d). We must now specify

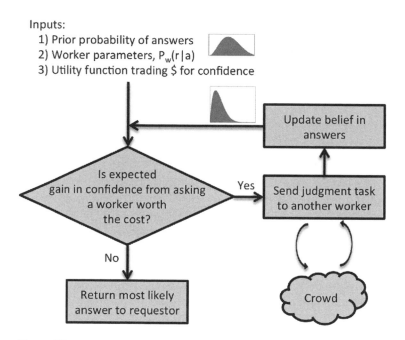

Figure 5.2
The POMDP model for a binary-vote workflow repeatedly decides if it is cost effective to ask another worker or whether the incremental reduction of uncertainty doesn't justify the cost.

the POMDP observation model, which encodes the probability of various observations as a function of the (hidden) state—for example, the probability of a worker's answering Yes when the correct answer is No. This is just another name for the worker models we discussed above, and we can use the methods of Whitehill et al. (2009).

A final input to the POMDP is a reward model, which in our case means the utility (or the money equivalent) of obtaining the correct answer. For example, the requester may specify that a correct answer is worth \$1.00 and an incorrect result (e.g., answering No when the true hidden answer is Yes) is equivalent to a −\$1.00 penalty. This utility is important, since it allows the controller to trade off between cost and quality. If a correct answer is very important to the requester, the controller should enlist more workers. In essence, the POMDP controller can compare the cost of asking an additional worker (or multiple workers) against the expected utility gain due to increased likelihood of a correct answer, and then ask another worker only if the net marginal utility is positive.

The literature (Kaelbling, Littman, and Cassandra 1998; Poupart 2011) describes numerous algorithms for solving POMDPs, and many of them can be adapted for this problem. Dai et al. (2013) try lookahead search, discretized dynamic programming, and Upper Confidence Bounds Applied on Trees (UCT). The exact choice of algorithm is not important. Most algorithms will perform better than a static fixed number of instances per question or even a hand-coded policy such as "ask two workers; if they agree, return that answer, otherwise break the tie with a third vote." Dynamic control is superior because it automatically adapts to the difficulty of the questions and the accuracy of the workers, capturing some valuable insights. First, if a question is easy (that is, workers agree on its answer) the controller will not ask for many more judgments, but if workers disagree the question may benefit from more judgments. Second, if a worker who is known to be excellent answers a question the controller may not ask for more judgments. Finally, if a question is deemed very difficult (e.g., good workers disagree on its answer) the system may decide that finding the correct answer is too expensive. In such cases, the controller will quit early even if it is not very sure about the answer. These kinds of intelligent decisions make it quite valuable for requesters to use AI techniques for workflow optimization.

Other researchers have also studied this problem in slightly different settings. Parameswaran et al. (2010) investigated other budgeted optimization problems and their theoretical properties. Waterhouse (Waterhouse 2013) investigated information-theoretic measures and used information content of a judgment as its value. Lin, Mausam, and Weld (2012a) studied controllers for data-filling questions, in which a worker must choose from an unbounded number of possible answers to a question rather than from two or a small number of known choices.

Kamar, Hacker, and Horvitz (2012) studied the interesting problem of mixed initiative systems, in which a machine model provides its own judgment for each question (and perhaps an associated confidence). Responses from human workers may validate or challenge the machine's answer. However, not all questions may require a response from a worker. If the machine is very confident, the controller may choose to ask only few human workers or none at all. Kamar et al. developed modifications of the UCT algorithm for creating their controller that can enhance machine answers with workers' judgments for additional accuracy.

Along the same lines, Lin, Mausam, and Weld (2012b) used multiple kinds of evidence in the form of different workflows (or different ways of asking the same question), exploiting the observation that some questions may be best asked in one form and others in a different form. The system of Lin et al. can automatically switch between such workflows to dramatically improve the quality of the output without increasing the cost.

Another important aspect of control algorithms is model learning. So far we have assumed that the controller knows the probability distribution for question difficulty, the ability parameters of the workers, and the observation probabilities. In some cases a human designer can estimate those numbers, but fortunately it is possible for the controller to *learn* their values even as it is controlling the workflow. A powerful formalism for balancing model learning with reward optimization is *reinforcement learning* (Sutton and Barto 1998). In RL-based control, the controller is in charge from the start; it naturally shifts its focus from model learning (exploration) in the beginning to reward maximization (exploitation) later. Lin, Mausam, and Weld (2012b) have used RL and found it to yield results of equivalent quality without an explicit learning phase. Kamar, Kapoor, and Horvitz (2013) describe a similar approach applied to citizen science applications. We expect RL methods to become increasingly popular, since they dramatically reduce the entry barrier for the use of AI technology in crowdsourcing.

Selecting the Best Question to Ask

So far we have focused on the control of a single question when the agent's goal is to obtain a correct answer in a cost-efficient manner. However, requesters typically turn to crowdsourcing only when they have a large number of questions. An interesting decision problem concerning the selection of questions arises when workers are unreliable. Given a fixed budget, should one divide resources equally, asking the same number of workers to tackle each question? Might it be better to dynamically allocate workers to questions on the basis of their reliability? The best policy is a function of how the answers will be used.

One common scenario is to use crowdsourced responses as a set of labeled examples for training a machine learning classifier. The decision problem can be formalized as follows: Suppose we have a large number of unlabeled questions, $u_1, \ldots, u_n$. Moreover, suppose that we have already asked workers for judgments to some of the questions $q_1, \ldots, q_k$. For each question q_i we may have asked multiple workers and hence we may have an aggregate answer (and associated confidence) for use in training the classifier. The decision problem is "Which question do we pick next for (re-)labeling?" There are two competing sources of evidence for this decision: (1) An existing aggregate answer (labeled example) for a question may be inaccurate and may in fact hurt the classifier, suggesting that we may want to ask for another judgment (relabeling). (2) The classifier may be uncertain in some part of the hypothesis space, and we may want to pick a question on the basis of the classifier's uncertainty in its predictions about the unlabeled data (active learning).

Active learning, the problem of choosing which unlabeled example should next be given to the oracle, has been widely studied in the AI and ML literature (Settles 2012). However, before the advent of crowdsourcing little work had considered active learning with noisy labels and hence the possibility of relabeling. Recently, several strategies for addressing this problem have been explored. Sheng et al. (2008) and Ipeirotis et al. (2013) focus on how best to relabel already-labeled questions, comparing repeated round-robin selection, answer-entropy-based selection, and other strategies. Donmez and Carbonell (2008) focus on various two-worker scenarios. For instance, they develop an algorithm to determine the best examples to label and by whom they should be labeled if one worker is infallible but costly and the other worker is fallible but cheap. Wauthier and Jordan (2011) propose a general utility-theoretic formulation to evaluate the expected utility gain for each question and pick the one with maximum gain.

Lin, Mausam, and Weld (2014) approach the tradeoff from a different direction and instead consider conditions under which relabeling a small number of examples is better than labeling a large number of examples once. They find that the inductive bias of the classifier and the accuracy of workers have profound effects on which strategy results in higher accuracies. We expect increasing attention to be paid to this problem of active learning with noisy annotators in coming years.

Selecting the Best Worker for a Question

Since some workers are more skilled or less error prone than others, it can be useful to match workers and questions. Amazon Mechanical Turk is not an ideal platform for worker allocation, since it resembles a "pull" model in which the workers choose their next tasks. However, other platforms, such as Zooniverse (Lintott et al. 2008), implement a "push" model in which the system decides what questions to send to a worker.

Various desiderata compete for such an allocation. The budget or the time available may be limited, so assigning every question to the best worker may not be feasible. More generally, one usually wishes to allocate questions so as to achieve a high-quality result while ensuring worker satisfaction, which is especially important in citizen science and in other applications that make use of volunteer labor. This may imply an even distribution of questions across workers and across each worker's questions. Parameters governing these distributions represent additional learning problems.

Various attempts have been made to study this problem in restricted scenarios. For example, Karger, Oh, and Shah (2011a,b, 2013) provide performance guarantees for a global worker assignment, but disallow adaptive question assignment, which would enable learning about workers' skills over time to better utilize good workers. On the other hand, Chen, Lin, and Zhou (2013) learn workers' skills and adaptively select promising workers, but do not bound the total number of questions allowed per worker to ensure that no worker is overburdened.

Ho and Vaughan (2012), Ho, Jabbari, and Vaughan (2013), and Tran-Thanh et al. 2012) assume constraints on the number of questions that may be assigned to any single worker, and divide the control problem into two explicit phases of exploration and exploitation. Ho et al. study a scenario in which each worker has a hidden skill level for each question. Their model assumes that workers arrive randomly, one at a time. Tran-Thanh et al. allow the system to select workers, but assume that questions are uniform and that a single worker completes each question, after which the controller is informed of the quality of the job each worker performed. In most crowdsourcing settings, multiple workers are needed to ensure quality, and quality is not directly observable.

Others consider the problem of selecting workers in the context of active learning. Yan et al. (2011) adaptively select the most confident worker for a given question, using a model based on question features. Donmez, Carbonell, and Schneider (2009) facilitate a gradual transition from exploration (learning about worker parameters) to exploitation (selecting the best worker for a question) by using upper confidence intervals to model worker reliability. Both Yan et al. and Donmez et al. first select the question and then choose the appropriate worker. By contrast, Wauthier and Jordan (2011) design an algorithm for selecting both a question and a worker that is geared toward learning about latent parameters.

The majority of worker-selection methods seek to discover the best workers and use them exclusively, but in volunteer crowdsourcing it is crucial to assign appropriate questions to all workers regardless of their skill. Bragg et al. (2014) study the problem of routing questions in parallel to all available workers in cases in which the questions vary in difficulty and the workers vary in skill, and develop adaptive algorithms that provide maximal benefit when workers and questions are diverse.

Shahaf and Horvitz (2010) studied generalized task markets in which the abilities of various workers were known but workers charged different rates for their services. They studied worker selection given a task and desired utility so that the workers with appropriate skill levels and payment profiles were chosen for a given task. Zhang et al. (2012) took a different approach entirely and shifted the burden of finding the appropriate worker from the system to the workers, noting that workers themselves may be best equipped to locate another worker with the appropriate skills for completing a question. Overall, worker-question allocation is an exciting problem for which there does not yet exist a satisfactory solution. We expect considerable progress in the next few years, in view of the problem's importance.

Controlling Workflows for Complex Objectives

Most AI research on workflow control has focused on simple multiple-choice questions as a natural first step. But the true power of crowdsourcing will be realized when we optimize workflows for more complex tasks. A wide variety of complex tasks have already been explored within this framework. Examples include computing a max from a set of items (Guo, Parameswaran, and Garcia-Molina 2012), multi-label classification and generating a taxonomy of items (Bragg, Mausam, and Weld 2013), iterative improvement workflow for writing image descriptions (Dai, Mausam, and Weld 2010, 2011), creating plans for achieving a goal (Kaplan et al. 2013), and selecting between multiple alternative workflows for a given task (Lin, Mausam, and Weld 2012b). In most of these settings the general abstraction includes defining a state space that encapsulates the agent's current belief about progress toward the requester's objective, estimating the value of each possible task, issuing the best task, and repeating the process on the basis of the expected outcome and any new information observed.

The exact mechanism for computing the value of human actions depends on the high-level objective. In cases in which the exact POMDP can be solved, the POMDP policy is used to select the workers' tasks. In other cases, simpler strategies have been used to reduce the computation involved. For example, greedy action selection was used to guide multi-label classification (Bragg, Mausam, and Weld 2013), and limited lookahead search was used to control iterative improvement.

As a case study we will briefly discuss the iterative improvement workflow (figure 5.3) introduced by Little et al. (2010) and optimized by Dai et al. (2010, 2011, 2013). Iterative improvement has been used to accomplish a wide range of objectives, such as deciphering human handwriting (Little et al. 2010), but for concreteness we will discuss a case in which the objective is to generate high-quality English captions for images. The workflow starts by asking a worker to write an initial caption for the picture. Then, at each iteration, a group of workers are shown the image and the current caption and asked to improve the caption by smoothing the writing or adding details. Another group of workers are shown the two descriptions (original and "improvement") and asked to select the best caption. These votes are aggregated, and the best description is adopted for the next iteration.

From the point of view of AI control, there are three actions that can be performed during execution of this workflow:

Issue a question asking for another improvement.
Issue a ballot question, requesting another comparison vote.
Submit the current description.

To pose the control problem as a POMDP, we first define the world state: the qualities of the two image descriptions. Let's use q_1 to denote the quality of the base description and q_2 to denote the quality of the new "improved" description. If no improvement has yet been requested, q_2 is undefined. We can constrain the qualities to be real numbers in [0, 1], where 1 represents an idealized perfect description of the image and 0 denotes the worst imaginable description. Next, we define the POMDP actions corresponding to asking a worker to improve a

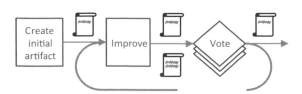

Figure 5.3
Control flow for an interactive improvement workflow (adapted from Little et al. 2010).

description or to compare two descriptions. The improvement model computes a probability distribution for possible values of $q_2 \in [0, 1]$ given that a worker with ability γ_{imp} tried to improve a description of quality q_1. Similarly, the voting model computes the probability that a worker of ability γ_{vote} will respond that description 1 is better when shown descriptions whose qualities are q_1 and q_2. Naturally, the probability of a mistake (saying "description one is better" when $q_2 > q_1$) increases if $|q_1 - q_2|$ is small and is inversely related to the worker's skill, γ_{vote}.

So far we have defined the dynamics of a POMDP. The final step is defining the utility function for the system to maximize, which will have two parts: the benefit due to returning a good description and the cost paid to workers. Clearly, the utility of returning a description with quality q should be a monotonically increasing function of q, though different requesters will assign different values to different qualities. Most people find it hard to articulate their precise utility function, and this has led to techniques for *preference elicitation*, which usually try to induce a general utility function from a small set of concrete judgments that are easier for people to answer (Boutilier 2002; Gajos and Weld 2005).

The definition of the POMDP is complete; now we have to solve the POMDP in order to produce a policy that says which action to execute as a function of the agent's beliefs. Because the state space of this POMDP is continuous (qualities are continuous variables), exact policy construction is extremely difficult. Dai et al. (2013) implemented this model and tried several approximate solution techniques, using supervised learning to induce the probabilistic transition functions from labeled training data. They found that POMDP-based control produced descriptions with the same quality as the original hand-coded policy with 30 percent less labor.

Even more interesting, the AI policy achieved its savings by issuing voting jobs in a dramatically asymmetrical fashion. In the hand-coded policy of Little et al. (2010), two workers always were asked to compare the original description and the "improved" description. If the assessments agreed, the two workers adopted the consensus description for the next iteration. If the two workers disagreed, the policy asked a third worker to break the tie. Thus, on average, the hand-coded policy issued about 2.5 voting questions per iteration. In contrast, the POMDP policy, illustrated in figure 5.4, issues *no* voting questions in the early iterations, which allows it to issue five or more voting questions in later iterations. In hindsight, this allocation makes sense—since it is

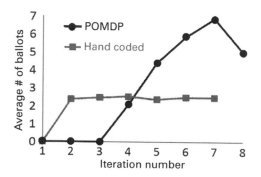

Figure 5.4
The POMDP controller for an iterative improvement workflow allocates resources differently than the hand-engineered policy, issuing more vote questions (ballots) in later iterations, when the comparisons are harder to make and additional opinions are needed for accurate decisions. Source: Dai et al. 2013.

relatively easy for workers to improve a description early on, there is little reason for workers to waste their time verifying the improvement's quality. After a few cycles of improvement, however, the description has become increasingly good and therefore harder and harder to improve. Now the POMDP chooses to spend additional resources issuing comparison votes, since it wants to be sure about which description to adopt for the next iteration.

This example points to the value of using AI control technology for complex tasks. For such tasks, often a human designer is unable to think through all possibilities, and hand-engineering a control policy that consistently exhibits optimal behavior can be difficult. Data-driven control approaches prove much more robust to corner cases and often end up saving large amounts of money. In other work, Bragg, Mausam, and Weld (2013) showed that they could categorize a large number of items into multiple categories with the same accuracy as hand-engineered policies while using less than 10 percent of the labor.

Unfortunately, the use of AI decision-theoretic methods is limited by its dependence on AI practitioners. The AI models and algorithms change somewhat depending on the task at hand. They require a level of mathematical sophistication that is often too great a barrier for typical requesters who are considering crowdsourcing. Weld, Mausam, and Dai (2011) sketched the architecture of a general-purpose system that will take a new workflow written in a high-level description language and automatically optimize it to control the workflow intelligently. If researchers can implement systems of this form and make them easy to use, AI methods may transform crowdsourcing practice in the years to come.

Minimizing Latency

Although most of the research on decision-theoretic control has focused on minimizing cost, there are a number of important situations in which latency—the time required to complete a task or a workflow—is especially important. For example, when crowdsourced workers are used to interpret a smartphone photo of a street sign for visually challenged users (Lasecki et al. 2013), a quick response is essential. Other low-latency applications include text editing (Bernstein et al. 2011), the selection of key frames from video, and captioning (Lasecki et al. 2012).

We may able to obtain near-instantaneous work if we pre-employ several workers so that when work arrives they are waiting to work on it (Bigham et al. 2010). This technique, termed the *retainer model*, has been studied analytically to determine the minimum number of retained workers required to achieve a quick response and the effect of delay on workers' attentiveness. Queuing theory may be used to model the probable arrival times for users' requests and thus the expected wait times for workers. We can then choose the number of workers to optimize the total cost subject to a constraint on expected delays or probability of missing a request (Bernstein et al. 2012). If there are several tasks that require retainers, the waiting workers can share them, and thus the costs of waiting can be amortized across tasks.

Conclusion

Crowdsourcing, a popular form of collective intelligence, has close connections to artificial intelligence. An increasing number of machine learning applications are trained with data produced by crowd annotation. Furthermore, many AI methods can be used to improve crowdsourcing. In particular, expectation maximization may be used to aggregate the results of multiple imprecise workers, learning workers' accuracies at the same time. Partially observable Markov decision processes and related decision-theoretic approaches may be used to optimize the types and the number of questions given to workers. Since these AI methods are a very active area of research, we expect to see even more powerful methods in the coming years.

Glossary

requester an entity assembling a crowd for an objective

objective what the requester is trying to accomplish

worker   an entity answering questions on a crowdsourcing platform

question   a category or type of work, e.g., image captioning or handwriting recognition

question   a single unit of work, e.g., a single image for an image captioning task. A task can have many individual questions.

response   what a worker returns when given a question (sometimes called a label)

answer   the true, objective answer to a question, when one exists (Ideally a worker's response will be the answer, but sometimes workers make mistakes.)

workflow   a set of tasks, usually interrelated, given to workers to perform. Some tasks may be performed automatically by programs, but most are given to human workers.

Acknowledgments

We appreciate many helpful conversations with Lydia Chilton, Peng Dai, Shih-Wen Huang, Stephen Jonany, and Andrey Kolobov. This work was supported by the WRF / TJ Cable Professorship, by Office of Naval Research grant N00014-12-1-0211, and by National Science Foundation grants IIS 1016713 and IIS 1016465.

References

Aldous, D. J. 1985. Exchangeability and related topics. In *École d'Été de Probabilités de Saint-Flour XIII 1983*, volume 1117 of *Lecture Notes in Mathematics*. Springer. 10.1007/BFb0099421.

Bernstein, M. S., J. Brandt, R. C. Miller, and D. R. Karger. 2011. Crowds in two seconds: Enabling realtime crowd-powered interfaces. In *Proceedings of the 24th Annual ACM Symposium on User Interface Software and Technology*. ACM.

Bernstein, M., D. Karger, R. Miller, and J. Brandt. 2012. Analytic methods for optimizing realtime crowdsourcing. Presented at Collective Intelligence Conference.

Bigham, J. P., C. Jayant, H. Ji, G. Little, A. Miller, R. C. Miller, R. Miller, et al. 2010. Vizwiz: Nearly real-time answers to visual questions. In *Proceedings of the 23nd Annual ACM Symposium on User Interface Software and Technology*. ACM.

Boutilier, C. 2002. A POMDP formulation of preference elicitation problems. Presented at AAAI Conference on Artificial Intelligence.

Bragg, J., and A. Kolobov. Mausam, and D. S. Weld. 2014. Parallel task routing for crowdsourcing. Presented at AAAI Conference on Human Computation and Crowdsourcing.

Bragg, J. Mausam, and D. S. Weld 2013. Crowdsourcing multi-label classification for taxonomy creation. Presented at Conference on Human Computation & Crowdsourcing.

Callison-Burch, C. 2009. Fast, cheap, and creative: Evaluating translation quality using Amazon's Mechanical Turk. In *Proceedings of the 2009 Conference on Empirical Methods in Natural Language Processing*, volume 1. Association for Computational Linguistics.

Chen, X., Q. Lin, and D. Zhou. 2013. Optimistic knowledge gradient policy for optimal budget allocation in crowdsourcing. Presented at International Conference on Machine Learning.

Chilton, Lydia B., Greg Little, Darren Edge, Daniel S. Weld, and James A. Landay. 2013. Cascade: Crowdsourcing taxonomy creation. In *Proceedings of the 2013 ACM Annual Conference on Human Factors in Computing Systems*. ACM.

Dai, P., C. H. Lin, Mausam, and D. S. Weld. 2013. POMDP-based control of workflows for crowdsourcing. *Artificial Intelligence* 202: 52–85.

Dai, P., Mausam, and D. S. Weld. 2010. Decision-theoretic control of crowd-sourced workflows. Presented at AAAI Conference on Artificial Intelligence.

Dai, P., Mausam, and D. S. Weld. 2011. Artificial intelligence for artificial artificial intelligence. Presented at AAAI Conference on Artificial Intelligence.

Dawid, A., and A. M. Skene. 1979. Maximum likelihood estimation of observer error-rates using the EM algorithm. *Applied Statistics* 28 (1): 20–28.

Dekel, O., and O. Shamir. 2009a. Good learners for evil teachers. Presented at International Conference on Machine Learning.

Dekel, O., and O. Shamir. 2009b. Vox populi: Collecting high-quality labels from a crowd. Presented at Annual Conference on Learning Theory.

Donmez, P., and J. G. Carbonell. 2008. Proactive learning: Cost-sensitive active learning with multiple imperfect oracles. In *Proceedings of the 17th ACM Conference on Information and Knowledge Management*. ACM.

Donmez, P., J. G. Carbonell, and J. Schneider. 2009. Efficiently learning the accuracy of labeling sources for selective sampling. In *Proceedings of the 15th ACM SIGKDD International Conference on Knowledge Discovery and Data Mining*. ACM.

Ferrucci, D. A., E. W. Brown, J. Chu-Carroll, J. Fan, D. Gondek, A. Kalyanpur, A. Lally, et al. 2010. Building Watson: An overview of the DeepQA project. *AI Magazine* 31 (3): 59–79.

Gajos, K., and D. S. Weld. 2005. Preference elicitation for interface optimization. In *Proceedings of the 18th Annual ACM Symposium on User Interface Software and Technology*. ACM.

Guo, S., A. G. Parameswaran, and H. Garcia-Molina. 2012. So who won? Dynamic max discovery with the crowd. In *Proceedings of the 2012 ACM SIGMOD International Conference on Management of Data*. ACM.

Ho, C.-J., and J. W. Vaughan. 2012. Online task assignment in crowdsourcing markets. Presented at AAAI Conference on Artificial Intelligence.

Ho, C.-J., S. Jabbari, and J. W. Vaughan. 2013. Adaptive task assignment for crowdsourced classification. Presented at International Conference on Machine Learning.

Horvitz, E. 2007. Reflections on challenges and promises of mixed-initiative interaction. *AI Magazine* 28 (2): 13–22.

Hsueh, P.-Y., P. Melville, and V. Sindhwani. 2009. Data quality from crowdsourcing: A study of annotation selection criteria. In *Proceedings of the NAACL HLT 2009 Workshop on Active Learning for Natural Language Processing*. Association for Computational Linguistics.

Ipeirotis, P. G., F. Provost, V. S. Sheng, and J. Wang. 2013. Repeated labeling using multiple noisy labelers. *Data Mining and Knowledge Discovery* (DOI 10.1007/s10618-013 -0306-1).

Kaelbling, L. P., M. L. Littman, and A. R. Cassandra. 1998. Planning and acting in partially observable stochastic domains. *Artificial Intelligence* 101 (1–2): 99–134.

Kajino, H., Y. Tsuboi, and H. Kashima. 2012. A convex formulation for learning from crowds. Presented at AAAI Conference on Artificial Intelligence.

Kamar, E., S. Hacker, and E. Horvitz. 2012. Combining human and machine intelligence in large-scale crowdsourcing. Presented at International Conference on Autonomous Agents and Multiagent Systems.

Kamar, E., A. Kapoor, and E. Horvitz. 2013. Lifelong learning for acquiring the wisdom of the crowd. In Proceedings of International Joint Conference on Artificial Intelligence.

Kaplan, H., I. Lotosh, T. Milo, and S. Novgorodov. 2013. Answering planning queries with the crowd. *Proceedings of the VLDB Endowment International Conference on Very Large Data Bases* 6 (9): 697–708.

Karger, D. R., S. Oh, and D. Shah. 2011a. Budget-optimal crowdsourcing using low-rank matrix approximations. Presented at conference on Communication, Control, and Computing.

Karger, D. R., S. Oh, and D. Shah. 2011b. Iterative learning for reliable crowd-sourcing systems. Presented at Neural Information Processing Systems conference.

Karger, D. R., S. Oh, and D. Shah. 2013. Efficient crowdsourcing for multi-class labeling. In *Proceedings of the ACM SIGMETRICS/International Conference on Measurement and Modeling of Computer Systems*. ACM.

Lasecki, W. S., C. D. Miller, A. Sadilek, A. Abumoussa, D. Borrello, R. S. Kushalnagar, and J. P. Bigham. 2012. Real-time captioning by groups of non-experts. In *Proceedings of the 25th Annual ACM Symposium on User Interface software and Technology*. ACM.

Lasecki, W. S., P. Thiha, Y. Zhong, E. L. Brady, and J. P. Bigham. 2013. Answering visual questions with conversational crowd assistants. In *Proceedings of the 15th International ACM SIGACCESS Conference on Computers and Accessibility*. ACM.

Lin, C. H. Mausam, and D. S. Weld. 2012a. Crowdsourcing control: Moving beyond multiple choice. In proceedings of Conference on Uncertainty in Artificial Intelligence.

Lin, C. H. Mausam, and D. S. Weld. 2012b. Dynamically switching between synergistic workflows for crowdsourcing. Presented at AAAI Conference on Artificial Intelligence.

Lin, C. H. Mausam, and D. S. Weld. 2014. To re(label), or not to re(label). In proceedings of AAAI Conference on Human Computation and Crowdsourcing.

Lintott, C., K. Schawinski, A. Slosar, K. Land, S. Bamford, D. Thomas, M. J. Raddick, et al. 2008. Galaxy zoo: Morphologies derived from visual inspection of galaxies from the Sloan digital sky survey. *Monthly Notices of the Royal Astronomical Society* 389 (3): 1179–1189.

Little, G., L. B. Chilton, M. Goldman, and R. C. Miller. 2010. TurKit: Human computation algorithms on Mechanical Turk. In *Proceedings of the 23nd Annual ACM Symposium on User Interface Software and Technology*. ACM.

Liu, Q., J. Peng, and A. Ihler. 2012. Variational inference for crowdsourcing. Presented at Neural Information Processing Systems conference.

Marcus, A., E. Wu, D. R. Karger, S. Madden, and R. C. Miller. 2011. Human-powered sorts and joins. *Proceedings of the VLDB Endowment* 5 (1): 13–24.

Mason, W. A., and D. J. Watts. 2009. Financial incentives and the "performance of crowds". *SIGKDD Explorations* 11 (2): 100–108.

Oleson, D., A. Sorokin, G. P. Laughlin, V. Hester, J. Le, and L. Biewald. 2011. Programmatic gold: Targeted and scalable quality assurance in crowdsourcing. In *Human Computation* (AAAI Workshops, WS-11-11).

Parameswaran, A., H. Garcia-Molina, H. Park, N. Polyzotis, A. Ramesh, and J. Widom. 2010. Crowdscreen: Algorithms for filtering data with humans. In Proceedings of 36th International Conference on Very Large Data Bases, Singapore.

Poupart, P. 2011. An introduction to fully and partially observable Markov decision processes. In *Decision Theory Models for Applications in Artificial Intelligence: Concepts and Solutions*, ed. E. Sucar, E. Morales and J. Hoey. IGI Global.

Prelec, D., and H. S. Seung. 2007. An algorithm that finds truth even if most people are wrong. http://www.eecs.harvard.edu/cs286r/courses/fall12/papers/Prelec10.pdf.

Raykar, V. C., S. Yu, L. H. Zhao, and G. Valadez. 2010. Learning from crowds. *Journal of Machine Learning Research* 11: 1297–1322.

Settles, B. 2012. *Active Learning: Synthesis Lectures on Artificial Intelligence and Machine Learning*. Morgan & Claypool.

Shahaf, D., and E. Horvitz. 2010. Generalized markets for human and machine computation. Presented at AAAI Conference on Artificial Intelligence.

Sheng, V. S., F. Provost, and P. G. Ipeirotis. 2008. Get another label? Improving data quality and data mining using multiple, noisy labelers. In *Proceedings of the Fourteenth ACM SIGKDD International Conference on Knowledge Discovery and Data Mining*. ACM.

Sheshadri, A., and M. Lease. 2013. Square: A benchmark for research on computing crowd consensus. In proceedings of AAAI Conference on Human Computation and Crowdsourcing.

Snow, R., B. O'Connor, D. Jurafsky, and A. Ng. 2008. Cheap and fast—but is it good? Evaluating non-expert annotations for natural language tasks. In *Proceedings of the Conference on Empirical Methods in Natural Language Processing*. ACM.

Sun, Y.-A., and Dance, C. R. 2012. When majority voting fails: Comparing quality assurance methods for noisy human computation environment. *CoRR* abs/1204.3516.

Sun, Y.-A., S. Roy, and G. Little. 2011. Beyond independent agreement: A tournament selection approach for quality assurance of human computation tasks. In *Human Computation: Papers from the 2011 AAAI Workshop*. AAAI Press.

Sutton, R. S., and A. G. Barto. 1998. *Reinforcement Learning*. MIT Press.

Thrun, S. 2006. A personal account of the development of Stanley, the robot that won the DARPA grand challenge. *AI Magazine* 27 (4): 69–82.

Tran-Thanh, L., S. Stein, A. Rogers, and N. R. Jennings. 2012. Efficient crowdsourcing of unknown experts using multi-armed bandits. Presented at European Conference on Artificial Intelligence.

Venetis, P., H. Garcia-Molina, K. Huang, and N. Polyzotis. 2012. Max algorithms in crowdsourcing environments. In *Proceedings of the 21st International Conference on World Wide Web*. ACM.

Waterhouse, T. P. 2013. Pay by the bit: An information-theoretic metric for collective human judgment. In *Proceedings of the 2013 Conference on Computer Supported Cooperative Work*. ACM.

Wauthier, F. L., and M. I. Jordan. 2011. Bayesian bias mitigation for crowdsourcing. Presented at Neural Information Processing Systems conference.

Weiss, G., ed. 2013. *Multiagent Systems*, second edition. MIT Press.

Weld, D. S. Mausam; and Dai, P. 2011. Human intelligence needs artificial intelligence. Presented at AAAI Conference on Human Computation and Crowdsourcing.

Welinder, P., S. Branson, S. Belongie, and P. Perona. 2010. The multidimensional wisdom of crowds. Presented at Neural Information Processing Systems conference.

Whitehill, J., P. Ruvolo, J. Bergsma, T. Wu, and J. Movellan. 2009. Whose vote should count more: Optimal integration of labels from labelers of unknown expertise. Presented at Neural Information Processing Systems conference.

Yan, Y., R. Rosales, G. Fung, and J. G. Dy. 2011. Active learning from crowds. Presented at International Conference on Machine Learning.

Zhang, H., E. Horvitz, Y. Chen, and D. C. Parkes. 2012. Task routing for prediction tasks. In *Proceedings of the 11th International Conference on Autonomous Agents and Multiagent Systems*, volume 2. International Foundation for Autonomous Agents and Multiagent Systems.

Cognitive Psychology

Editors' Introduction

Even though the field of collective intelligence sees intelligence in unusual places, any comprehensive survey of the field should certainly include the first place where we humans saw something we called "intelligence": our own minds. And although philosophers and others have been discussing various aspects of human intelligence since at least the time of Aristotle, scientific study of the phenomenon began only about 100 years ago. It has proceeded along two mostly separate paths.

One path involves measuring the differences among individuals in their mental functioning. For instance, Spearman (1904) was the first to demonstrate statistically the non-intuitive fact that there is a single statistical factor for an individual that predicts how well the individual will perform a very wide range of mental tasks. This factor is very similar to what we usually call "intelligence," and it is what all modern IQ tests measure (Deary 2000).

The other path involves studying the mental processes that make intelligent behavior possible. Though there is not complete consensus on how to categorize these cognitive processes, a typical list (see, e.g., Reisberg 2013; Frankish and Ramsey 2012) would include perception, attention, memory, learning, decision making, and problem solving.

As an example of the second path, it is well known that there are at least two kinds of human memory. Working memory decays rapidly without conscious attention, stores verbal and visual information separately, and has very limited capacity (Atkinson and Shiffrin 1968; Baddely and Hitch 1974). Long-term memory can store much larger amounts of information for much longer times and depends strongly, for both storage and retrieval, on the meaning of the information being remembered (see, e.g., Craik and Lockart 1972).

As another example, psychologists developed the concept of "learning curves" to describe how humans often improve their performance on new tasks rapidly at first and then more slowly (see, e.g., Ebbinghaus 1885; Thorndike 1898). Later, economists and organization theorists generalized that concept to learning by other kinds of collectively intelligent systems, such as companies (Benkard 2000; Hirshleifer 1962; Argote 1999; March 1991).

From the perspective of collective intelligence, many intriguing questions involve whether—and how—these cognitive processes are manifested, not just in individual human brains, but also in groups of individuals—whether those individuals are people, computers, insects, or companies. For instance, do ant colonies (as described in the chapter on biology) have equivalents of long-term memory and working memory at the level of the whole colony? Or, as we'll examine in the chapter on organizational behavior, how do human organizations demonstrate collective cognitive processes such as group memory, group learning, and group decision making?

In the chapter that follows, Steyvers and Miller focus on another way of bringing together an understanding of individual and group cognitive processes. Their focus is on the kind of group decision making that gives rise to the "wisdom of crowds" effect. In this kind of decision making, each member of a group estimates a quantity or selects an answer to a question, and the group's decision is a function (often the average) of the individual answers. The resulting group answer is often surprisingly accurate. Steyvers and Miller provide a comprehensive analysis of the individual cognitive factors that affect a group's decision making in such a situation, and they show that things are more complex than they might at first seem.

In the chapter by Steyvers and Miller, and in the field of psychology in general, the primary kind of evidence brought to bear on questions about intelligence comes from behavioral studies of humans. Randomized, controlled experiments are the "gold standard" for evidence, and statistical hypothesis testing is necessary to separate pattern from randomness. Psychology focuses more on behavioral evidence about what people *actually do* than on formal models (common in economics) about what people *should* do. Although some of the lessons psychologists have learned from these studies apply only to how intelligence works in individual human brains, many others illuminate aspects of how collective intelligence can arise in many other kinds of systems.

References

Argote, L. 1999. *Organizational Learning: Creating, Retaining and Transferring Knowledge.* Kluwer.

Atkinson, R. C., and R. M. Shiffrin. 1968. Human memory: A proposed system and its control processes. In *The Psychology of Learning and Motivation: Advances in Research and Theory*, ed. K. W. Spence. Academic Press.

Baddeley, A. D., and G. J. Hitch. 1974. Working memory. In *The Psychology of Learning and Motivation*, volume 8, ed. G. Bower. Academic Press.

Benkard, C. L. 2000. Learning and forgetting: The dynamics of aircraft production. *American Economic Review* 90 (4): 1034–1054.

Craik, F. I. M., and R. S. Lockhart. 1972. Levels of processing: A framework for memory research. *Journal of Verbal Learning and Verbal Behavior* 11: 671–684.

Deary, Ian J. 2000. *Looking Down on Human Intelligence.* Oxford University Press.

Ebbinghaus, H. 1885. *Memory: A Contribution to Experimental Psychology.* Dover.

Frankish, Keith, and William Ramsey. 2012. *The Cambridge Handbook of Cognitive Science.* Cambridge University Press.

Hirshleifer, J. 1962. The firm's cost function: A successful reconstruction? *Journal of Business* 35 (3): 235–255.

March, J. G. 1991. Exploration and exploitation in organizational learning. *Organization Science* 2 (1): 71–87.

Reisberg, Daniel. 2013. *The Oxford Handbook of Cognitive Psychology.* Oxford University Press.

Spearman, Charles. 1904. "General intelligence," objectively determined and measured. *American Journal of Psychology* 15 (2): 201–292.

Thorndike, E. L. 1898. Animal intelligence: An experimental study of the associative processes in animals. *Psychological Monographs* 2 (4): i–109.

Recommended Readings

Daniel Kahneman. 2011. *Thinking, Fast and Slow.* Farrar, Straus and Giroux.

This popular book, written for a general audience by a cognitive psychologist who received a Nobel Prize in economics, provides a comprehensive summary of recent work on how human minds actually work. According to Kahneman, human minds have two modes of operation. System 1 ("thinking fast") quickly gives intuitive answers to many questions that are good enough to be useful in many situations but are sometimes wrong in systematic ways. System 2 ("thinking slow") provides more carefully thought out logical answers that are usually more accurate but also more time consuming and difficult to produce. A question for readers of the current volume to contemplate is whether there are analogues of System 1 and System 2 in various kinds of collective intelligence.

Daniel Reisberg. 2013. *Cognition*, fifth edition. Norton.

Edward E. Smith and Stephen M. Kosslyn. 2007. *Cognitive Psychology: Mind and Brain*. Pearson Prentice-Hall.

These two introductory textbooks on cognitive psychology provide a good overview of what psychologists know about how human minds carry out the various processes involved in intelligent behavior.

James Surowiecki. 2004. *The Wisdom of Crowds*. Random House.

This book, which helped popularize the notion of collective intelligence, provides many provocative examples of how large groups of ordinary people can sometimes be smarter than a few experts. It opens with the now-classic example of the "wisdom of crowds" effect: averaging many people's guesses of the weight of an ox often provides a very accurate estimate.

P. E. Tetlock, B. A. Mellers, N. Rohrbaugh, and E. Chen. 2014. Forecasting tournaments: Tools for increasing transparency and improving the quality of debate. *Current Directions in Psychological Science* 23 (4): 290–295.

This article gives an overview of the Good Judgment Project, the winner of a massive tournament the object of which was to predict geopolitical events. The Good Judgment Project uses a number of today's state-of-the-art techniques for group forecasting, including many of the techniques described in the chapter you are about to read.

Cognition and Collective Intelligence

Mark Steyvers and Brent Miller

Cognitive and psychological research provides useful theoretical perspectives for understanding what is happening inside the mind of an individual in tasks such as memory recall, judgment, decision making, and problem solving—including meta-cognitive tasks, in which individuals reflect on their own performance or that of others. Although certain types of interactions between group members can allow groups to collectively process information (e.g., transactive memory; see Wegner 1987) or to utilize shared mental states through patterns in the environment (Norman 1993), the focus of this chapter will be on cognitive processes contained wholly within single minds that can affect group behavior.

We humans differ from many other collectively intelligent organisms in groups (see the chapter on human–computer interaction) in that we are measurably intelligent independent of one another. Understanding the cognitive processes within individuals can help us understand under what conditions collective intelligence might form for a group and how we might optimize that group's collective performance. These components, alone or in concert, can be understood to form the basic building blocks of group collective intelligence.

Consider the classic estimation task in which a group of individuals must determine the number of marbles in a jar. In the simplest conceptualization of this task, each individual independently provides an estimate and a statistical average of the estimates is taken as the crowd's answer. The statistical aggregate over individuals can often lead to an answer that is better than that arrived at by most of the individuals. This has come to be known as the "wisdom of crowds" effect (Ariely et al. 2000; Davis-Stober et al. 2014; Surowiecki 2004; Wallsten et al. 1997). Given simple, idealized tasks, it would appear that extracting the collective intelligence from a group of individuals merely requires choosing a suitable statistical aggregation procedure—no psychology or under-standing of the underlying cognitive processes is necessary. However, if we start to make more realistic assumptions about the estimation task or change it to make it more like complex, real-world situations, it quickly becomes obvious how psychological factors can come into play. Suppose, for example, that individuals give judgments that are systematically biased (e.g., they may overestimate the number of marbles in a jar because the marbles differ in size). How can we know what the potential biases are, and how to correct for them? Suppose that some individuals are better at a task than others, or do not understand the task, or aren't even paying attention. How do we identify the judgments that are more accurate? What are the measures that we can use to identify experts? If individuals share information about their judgments and their reasoning, how does the sharing affect the results? To fully understand how collective intelligence arises from a group of individuals, and how a group's collective wisdom can be improved, it is necessary to consider what is going on inside the human mind.

In this chapter, we will review the cognitive and psychological research related to collective intelligence. We will begin by exploring how cognitive biases can affect collective behavior, both in individuals and in groups. Next we will discuss expertise and consider how more knowledgeable individuals may behave differently and how they can be identified. We will also review some recent research on consensus-based models and meta-cognitive models that identify knowledgeable individuals in the absence of any ground truth. We will then look at how the sharing of information by individuals affects the collective performance, and review a number of studies that manipulate how information is shared. Finally, we will look at collective intelligence within a single mind.

Identifying and Correcting for Biases

Whether or not a group collectively arrives at sensible judgments depends largely on whether the individuals are sensible. The literature provides many examples of cognitive biases that can systematically distort the judgment of individuals (see, e.g., Hogarth 1975; Kahneman et al. 1982). For example, human probabilistic judgments may be over-confident about reported probabilities, may neglect the event's base rate, or may be biased by the desirability of the outcomes (Kahneman and Tversky 2000; Gilovich et al. 2002; Massey et al. 2011). Biased misperceptions of likelihood can have deleterious effects on entire economies (Taleb 2007). Conversely, individuals may often be sensitive to extraneous information that can be irrelevant to the judgment task at hand (Goldstein and Gigerenzer 2002). Systematic distortions that affect individuals' judgments can also affect group performance. Although uncorrelated errors at the level of individual judgments can be expected to average out in the group, systematic biases and distortions cannot be averaged out by using standard statistical averaging approaches (Simmons et al. 2011; Steyvers et al. 2014).

It has been shown to be possible to get subjects to debias their own estimates, at least to a degree, by training individuals in the potential biases of estimation (Mellers et al. 2014). Alternatively, it is possible, by understanding what these cognitive biases are, to correct them before performing statistical aggregation. In some domains, such as predicting the likelihood of low-probability events, subjects are systematically overconfident (Christensen-Szalanski and Bushyhead 1981). In judging other events that occur more frequently, as in weather forecasting, experts have more opportunity to properly calibrate their responses (Wallsten and Budescu 1983). When expert judgments are tracked over a period of time, it is possible to learn and correct for systematic biases. Turner et al. (2014) used hierarchical Bayesian models to learn a recalibration function for each forecaster. The calibrated individual estimates were then combined using traditional statistical methods, and the resulting aggregation was found to be more accurate than aggregates of non-calibrated judgments. Satopää et al. (2014) have proposed similar recalibration methods that shift the final group estimates, using either weighted or unweighted aggregation of the individual responses.

Human judgment can also be error-prone and inconsistent when information between interrelated events needs to be connected. For example, when people judge the likelihood of events that are dependent on one another, the result can lead to incoherent probability judgments that do not follow the rules of probability theory (Wang et al. 2011). Probabilities for interrelated events are coherent when they satisfy the axioms of probability theory. For example, the probability of a conjunction of events (A and B) has to be equal to or less than the probability of the individual events (A or B). However, people may not always connect these interrelations in logical ways and may fail to produce coherent probability judgments. Failure of coherence can occur at the individual level (Mandel 2005), but also can occur at the aggregate level in prediction markets (Lee et al. 2009). Similarly, probability judgments that are incoherent at the individual level cannot be expected to become coherent by averaging across individuals (Wang et al. 2011). Incoherence may persist even in the presence of financial incentives (Lee et al. 2009). Wang et al. (2011) proposed a weighted coherentization approach that combines credibility weighting with coherentization so that the aggregate judgments are guaranteed to obey the rules of probability; for instance, when they asked participants to forecast the outcome of the 2008 U.S. presidential election, some of the questions were about elementary events but others involved negations, conditionals, disjunctions, and conjunctions (e.g., "What is the probability that Obama wins Vermont and McCain wins Texas?"). Sometimes humans make errors in estimation when the task environment encourages them to do so. In a competitive environment with information sharing, there may be an advantage to not giving one's best estimates to others. Ottaviani and Sørensen (2006) studied professional financial forecasters and found that the incentive to distinguish oneself from one's fellow forecasters outweighed the traditional goal of minimizing estimation error. Depending on the nature of the competition, fairly complex cognitive strategies can be employed to generate answers that are not representative of individuals' true estimates. On the televised game show *The Price is Right*, contestants bid in sequential order on the price of an item, and the winner is the contestant who comes closest to the price without exceeding it. Contestants often give estimates that are quite far below the actual price (and presumably quite far from what they believe that price to be) in order to increase their odds of winning. Aggregation approaches that model the strategic considerations of such competitive environments and attempt to aggregate over inferred beliefs outperform standard aggregation methods (Lee et al. 2011). When competition is employed, a winner-take-all format with minimal information may be best suited to get the most useful estimates from individuals; there is reason to believe that people will be more likely to employ any unique information they may have to make riskier but more informative estimates for aggregation (Lichtendahl et al. 2013).

Identifying Expert Judgments

The presence of experts in a group can have a significant effect on the accuracy and behavior of collective intelligence. The ability to identify and use these experts is an important application in a wide range of real-world settings. Society expects experts to provide more qualified and accurate judgments within their domain of expertise (Burgman et al. 2011). In some domains, such as weather forecasting, self-proclaimed experts are highly accurate (Wallsten and Budescu 1983). However, self-identified or peer-assessed expertise may not always be a reliable predictor of performance (Tetlock 2005; Burgman et al. 2011). Expertise isn't always easy to identify—it can be defined in a number of different ways, including experience, qualifications, performance on knowledge tests, and behavioral characteristics (Shanteau et al. 2002). Procedures to identify experts can lead to mathematical combination approaches that favor better, wiser, more expert judgments when judgments from multiple experts are available (French 1985, 2011; Budescu and Rantilla 2000; Aspinall 2010; Wang et al. 2011). In the subsections that follow, we discuss a number of general approaches that have been developed to assess the relative expertise in weighted averages and model-based aggregation procedures.

Performance weighting
A classic approach to aggregate expert opinions is based on Cooke's method (Cooke 1991; Bedford and Cooke 2001; Aspinall 2010). Cooke's method requires an independent stand-alone set of seed questions (sometimes referred to as calibration or control questions) with answers known to the aggregator but unknown to the experts. On the basis of performance on these seed questions, weights are derived that can be used to up-weight or down-weight experts' opinions on the remaining questions that don't have known answers (at least at the time of the experiment). Aspinall (2010) gives several real-world examples of Cooke's method, such as estimating failure times for dams exposed to leaks. Previous evaluations of Cooke's method may have led to over-optimistic results because the same set of seed questions used to calculate the performance weights were also used to evaluate model performance (Lin and Cheng 2009). Using a cross-validation procedure, Lin and Cheng (ibid.) showed that the performance-weighted average and an unweighted linear opinion pool in which all experts were equally weighted performed about the same. They concluded that it wasn't clear whether the cost of generating and evaluating seed questions was justifiable. Liu et al. (2013) performed a theoretical analysis in a scenario in which the total number of questions that could be asked of judges was limited (e.g., each judge could estimate only fifty quantities), so that any introduction of seed questions necessarily reduced the number of questions with unknown ground truth (the questions of

ultimate interest). They found that under some conditions a small number of seed questions sufficed to evaluate the relative expertise of judges and measure any systematic response biases.

Using performance weighting, Budescu and Chen (2014) developed a contribution-weighted model in which the goal was to weight individuals by their contribution to the crowd in terms of the difference of the predictive accuracy of the crowd's aggregate estimate with, and without the judge's estimate in a series of forecasting questions. Therefore, individuals with a high contribution were those for which group performance would suffer if their judgment were omitted from the group average.

Generally, performance-based methods have the disadvantage that it can take time to construct seed questions with a known answer. As Shanteau et al. (2002) argued, experts may be needed in exactly those situations in which correct answers are not readily available. In forecasting situations, obvious choices for seed questions include forecasting questions that resolve during the time period over which the judge is evaluated. However, such procedures require an extended time commitment from judges and thus may not be practical in some scenarios.

Subjective Confidence

Another approach is to weight judgments by the subjective confidence expressed by the judges. In many domains, subjective confidence often demonstrates relatively low correlation with performance and accuracy (see, e.g., Tversky and Kahneman 1974; Mabe and West 1982; Stankov and Crawford 1997; Lee et al. 2012). However, in some cases a judge's confidence can be a valid predictor of accuracy. For example, in a group consisting of two people, a simple strategy of selecting the judgment of the more confident person (Koriat 2012) leads to better performance than relying on any single judgment. Koriat argues that subjective confidence may be driven more by common knowledge than by the correctness of the answer. However, it is possible to set up tasks in which the popular answer, typically associated with high confidence, is also the incorrect answer (Prelec and Seung 2006). Overall, performance from confidence-weighted judgments will depend heavily on the nature of the task and the degree to which the task is a representative sample of individuals (Hertwig 2012).

Coherence and Consistency

Coherence in probability judgments can be taken as a plausible measure of a judge's competence in probability and logic. Wang et al. (2011) and

Olson and Karvetski (2013) showed that down-weighting judgments of individuals associated with less coherent judgments (across questions) was effective in forecasting election outcomes. A related idea is that experts should produce judgments that are consistent over time such that similar responses are given to similar stimuli (Einhorn 1972, 1974). The within-person reliability or consistency can be used as a proxy for expertise, especially when it is combined with other cues for expertise such as discrimination (Shanteau et al. 2002; Weiss and Shanteau 2003; Weiss et al. 2009). One potential problem is that consistency is often assessed over short time intervals and with stimuli that are relatively easy to remember. In these cases, memory-retrieval strategies may limit the usefulness of consistency measures. Miller and Steyvers (2014) studied cases involving judgments that were difficult to remember explicitly and showed that consistency across repeated problems was strongly correlated with accuracy and that a consistency-weighted average of judgments was an effective aggregation strategy that outperformed the unweighted average.

Consensus-Based Models

The idea behind consensus-based models is that in many tasks the central tendency of a group leads to accurate answers. This group answer can be used as an estimate of the true answer to score individual members of a group. Individuals who produce judgments that are closer to the group's central tendency (across several questions) can be assumed to be more knowledgeable. Consensus-based models can therefore be used to estimate the knowledge of individuals in the absence of a known ground truth.

Consensus measures have been used in weighted averages where the judgments from consensus-agreeing individuals are up-weighted (Shanteau et al. 2002; Wang et al. 2011). Comprehensive probabilistic models for consensus-based aggregation were developed in the context of cultural consensus theory (Romney et al. 1987; Batchelder and Romney 1988) as well as in the context of observer-error models (Dawid and Skene 1979). To understand the basic approach, consider a scenario in which an observer has to figure out how to grade a multiple-choice test for which the answer key is missing. A consensus model posits a generative process in which each test taker, for each question, gives an answer that is a sample taken from a distribution in which the mean is centered on the latent answer key and the variance is treated as a variable that relates inversely to the latent ability of the observer.

Probabilistic inference can be used to simultaneously infer the answer key and the ability of each individual. Test takers with high ability are closer to the answer key on average; test takers with lower ability tend to deviate more from the answer key and from their higher-ability compatriots.

This consensus-based approach is not limited to problems for which the responses are discrete; it can also be used to estimate group responses over a continuous range of potential answers (Batchelder and Anders 2012). Consensus-based models are also able to account for variations in the difficulty of the questions. Consensus-based methods have led to many statistical models for crowdsourcing applications in which workers provide subjective labels for simple stimuli such as images (see, e.g., Smyth et al. 1995; Karger et al. 2011). Hierarchical Bayesian extensions have been proposed by Lipscomb et al. (1998) and by Albert et al. (2012).

Recently, consensus-based aggregation models have been applied to more complex decision tasks, such as ranking data (Lee et al. 2012, 2014). For example, individuals ranked a number of U.S. presidents in chronological order, or cities by their number of inhabitants. A simple generative model was proposed in which the observed ranking was based on the ordering of samples from distributions centered on the true answer but with variances determined by latent expertise levels. Lee et al. (2012) showed that the expertise levels inferred by the model were better correlated with actual performance than subjective confidence ratings provided by the participants.

Generally, consensus-based methods perform well on tasks that individuals do reasonably well (Weiss et al. 2009). One potential weakness of consensus-based methods is that they are vulnerable to cases in which agreement arises for reasons other than expertise. This can occur in challenging tasks on which the majority of individuals use heuristics. For example, when predicting the outcomes of certain sports tournaments, individuals who do not closely follow those tournaments may adopt heuristic strategies based on the familiarity of the teams (see, e.g., Goldstein and Gigerenzer 2002). Another potential issue is that in some cases it may be inappropriate to assume that there is a single latent answer or opinion for the whole group—there may be multiple clusters of individuals with divergent beliefs. In this case, consensus-based models must be extended to make inference over multiple groups with multiple answer keys; there has been preliminary work that shows that this may indeed be feasible (Anders and Batchelder 2012).

The Role of Meta-Cognition

The Bayesian Truth Serum (BTS) proposed by Prelec (2004) is an idea that incorporates metaknowledge—the knowledge of other people's judgments in aggregation. The BTS method was designed as an incentive mechanism to encourage truthful reporting. It can elicit honest probabilistic judgments even in situations in which the objective truth is difficult to obtain or intrinsically unknowable. It has been used to encourage people to answer survey questions truthfully (Weaver and Prelec 2013) and to estimate the prevalence of questionable research practices (John et al. 2012). However, it has also been tested in preliminary experiments on general knowledge questions (Prelec and Seung 2006) with which the performance of the method can be assessed objectively—for example, whether Chicago is the capital of Illinois. A minority of respondents might be expected to know the correct answer to that question. The majority of respondents might use simple heuristics that would lead them to the plausible yet incorrect answer. In the BTS approach, judges provide a private answer to a binary question and an estimate of the percentage of people who would give each response. The latter estimate involves metacognitive knowledge of other people. For each judge, a BTS score is calculated that combines the accuracy of the metacognitive judgments (rewarding an accurate prediction of other people's responses) and an information score that rewards surprisingly common responses. In the previous question, the correct answer—Springfield—will receive a high score if more people actually produced that answer than was predicted (metacognitively). Prelec and Seung (2006) showed that the BTS-weighted aggregate outperformed majority voting in a number of cases—cases in which only a minority of judges knew the correct answer. Though these initial empirical results are promising, it isn't clear how the BTS method will perform in areas, such as forecasting, in which the true answer isn't knowable at the time the question is asked, and meta-cognition about other people's forecasts might be biased in a number of ways. Recent research has also suggested that the metacognitive efficacy of individuals is positively correlated with the group's overall collective intelligence ability (Engel et al. 2014).

The Role of Information Sharing

Until recently, much of the work that has been done in collective decision making has involved a good deal of dynamic interaction among group members (see, e.g., Lorge et al. 1958). Often a group of properly trained people with a lot of experience in working together can make judgments that are more accurate than those of any of the individual members (Watson et al. 1991; also see the chapters in this volume on organizational behavior and on law and other disciplines). When members of a group haven't been specifically trained to work together, the results can be far more varied; group members may have difficulty coordinating their responses to obtain a consensus (Steiner 1972; Lorenz et al. 2011) and are more vulnerable to cognitive biases and errors (Janis 1972; Stasser and Titus 1987; Kerr et al. 1996). It has been suggested that groups in which the individuals interact are most effective when their collective decision is arrived at by a weighted average of each member's opinions (Libby et al. 1987).

One popular method for soliciting group judgments is the Delphi method (Rowe and Wright 1999). By separating individuals, having them individually answer guided questionnaires, and allowing them to view one another's responses and to provide updated feedback, the Delphi method allows individuals to weight their own expertise in relationship to others and (ideally) to provide better-informed estimates. These individual estimates are then combined via statistical aggregation similar to the previous methods discussed. As with the training of specialized decision-making groups, there is still a large cost associated with setting up and coordinating a Delphi-based decision process. There are a number of additional schemes for limited information sharing that avoid many of the social and cognitive biasing that is inherent in dynamic group decision making (Gallupe et al. 1991; Olson et al. 2001; Whitworth et al. 2001; Rains 2005—also see the chapters in this volume on human–computer interaction and artificial intelligence).

The effect of information sharing is strongly dependent on the type of network structure in which participants share information with one another (Kearns et al. 2006, 2012; Mason et al. 2008; Judd et al. 2010; Bernstein et al. 2011). For example, Mason et al. (2008) studied problem-solving tasks in which participants (corresponding to nodes on a network) were arranged in a number of different networks, some of them fully connected, some of them lattices, some of them random, and some of them small-world networks. The task for participants was to find the maximum of a continuous function with one input variable. Participants could probe the function with numerical values for the input variable and obtain feedback by the value returned by the function. The function was sufficiently complicated with multiple local modes such that no individual could cover the space of possibilities within a reasonable amount of time. Participants received information about their neighbors' guesses and outcomes. The results showed that the network configuration had a strong effect on overall performance. Individuals found good global solutions more quickly in the small-world networks, relative to lattices and random networks, presumably because information can spread very quickly in small-world networks. It is not entirely clear why the small-world networks performed better than the fully connected networks, however. In a fully connected network, participants have full information about all other participants, and theoretically they should be able to benefit from that information. Mason et al. (2008) proposed that "less is more" in small-world networks—that participants may be better able to pay attention to the information from a smaller number of neighbors.

Kearns et al. (2006) and Judd et al. (2010) studied decentralized coordination games on networks in which each participant solved only a small part of a global problem. In contrast with the study by Mason et al. (2008), individuals were required to coordinate their efforts in order to collectively produce a good global solution. One coordination game involved a coloring problem in which each participant was required to choose a color from a fixed set of colors that was different from those of his or her neighbors. The results showed that the network structure had a strong influence on solution times. Long-distance connections hurt performance on the coloring task. On the other hand, if the task was altered so that consensus solutions were rewarded (i.e., all nodes had the same color), the long-distance connections improved performance. Across many of these coordination tasks, performance of human subjects came close to the optimal solution. Kearns et al. (2012) reported that 88 percent of the potential rewards available to human subjects were collected.

Task sharing can also be beneficial when individuals must explore a large problem space to find good solutions. Khatib et al. (2011) used a collective problem-solving approach to scientific discovery to optimize protein folding. Each player manipulated the protein folding to find stable configurations. One group of participants found a breakthrough solution to the problem that then was adopted by other participants as a new starting point for their own solutions. Collaboration also makes sense when questions are sufficiently complex that subjects may have different parts of the answer (Malone et al. 2010). Miller and Steyvers (2011) explored rank-ordering tasks; the first subject in the task was given a random list ordering, and then each subject received the final ordering of a previous participant in an iterative fashion. In contrast with simpler information-passing tasks (see Beppu and Griffiths 2009), answers didn't necessarily converge on the correct ordering, but by aggregating across all subjects in the sequence it was possible to combine the partial knowledge of individuals into a nearly complete whole. Tools exist whereby this information sharing can be utilized to explicitly create external, shared collective knowledge as an aid to the collaborative process (Ren and Argote 2011). It has been shown that subjects are more susceptible to memory bias when given the responses of other subjects, but this can be overcome by using aggregation (Ditta and Steyvers 2013).

Collective Intelligence within Individuals

Whereas collective intelligence is often considered at the level of groups, we can also consider collective intelligence within an individual. In one experiment, Vul and Pashler (2008) asked individuals to estimate quantities (e.g., "What percentage of the world's airports are in the United States?") multiple times at varying time intervals. They found a "wisdom of the crowd" effect within one mind—the average of two guesses (from the same person) was more accurate than either of the individual guesses. This effect was larger if more time elapsed between the two estimates, presumably because participants' answers were less correlated because of strategic or memory effects. Hourihan and Benjamin (2010) found that the average of two guesses from individuals with short working-memory spans was more accurate than the average of two guesses from individuals with long working-memory spans, which suggested that the ability to remember the first response (as opposed to reconstructing an answer from general knowledge) might be an impediment to the "wisdom within one mind" effect.

Rauhut and Lorenz (2011) generalized Vul and Pashler's finding and demonstrated that the average over five repeated estimates was significantly better than the average from two repeated estimates (or a single estimate). This is somewhat surprising—one might assume that the first guess would already be based on all available information and that the subsequent guesses would not provide additional information. These findings show that there is an independent error component in the estimates that can be canceled by averaging. Generally, these findings also support the concept that subjective estimates arise as samples from probabilistic representations underlying perceptual and cognitive models (Gigerenzer et al. 1991; Fiser et al. 2010; Griffiths et al. 2012).

The exact procedure used to elicit repeated judgments has been found to influence the "wisdom within one mind" effect. For example, a method known as dialectical bootstrapping (Herzog and Hertwig 2009) is designed to facilitate the retrieval of independent information from memory. Participants are told that their first estimate is off the mark and are asked to consider knowledge that was previously overlooked, ignored, or deemed inconsistent with current beliefs. Herzog and Hertwig showed that dialectical bootstrapping led to higher accuracy than standard instructions. In the More-Or-Less-Elicitation (MOLE) method (Welsh et al. 2009), participants are asked to make repeated relative judgments in which they have to select which of two options they think is closer to the true value. The advantage of this procedure is that it avoids asking the exact same question, which might elicit an identical answer.

Discussion

Human cognition plays a major role in the formation of collective intelligence by groups. In order to understand the collective intelligence of groups, we need to understand how judgments made by individual minds are affected by errors, biases, strategies, and task considerations. By developing aggregation methods and models that correct for these factors, and by using debiasing procedures in which individuals are trained to avoid such mistakes, it is possible to make more intelligent collective decisions.

It is also necessary to understand how the collective performance of a group is affected by the group's composition, by the relative expertise of the members, and by the sharing of information (if there is any such sharing) by the members. Such an understanding can help us to identify individuals who tend to produce more accurate judgments and also can help us to determine how and when to allow individuals to share information so as to make better collective estimates. In addition, by understanding the meta-cognition of the individuals in a group—their understanding of the other individuals—we can learn more about an individual's knowledge than we can learn from that individual's judgments alone.

Finally, one of the most important roles for cognitive research is to further our understanding of individuals' mental representations that are used to produce judgments. Converging evidence suggests that human knowledge is inherently probabilistic. Not only does this affect how individuals retrieve information from themselves; it also affects how they view others' information. The nature of these mental representations has implications for the kinds of aggregation models that are effective in combining human judgments, and for how collective intelligence arises generally.

References

Albert, I., S. Donnet, C. Guihenneuc-Jouyaux, S. Low-Choy, K. Mengersen, and J. Rousseau. J. 2012. Combining expert opinions in prior elicitation. *Bayesian Analysis* 7 (3): 503–532.

Anders, R., and W. H. Batchelder. 2012. Cultural consensus theory for multiple consensus truths. *Journal of Mathematical Psychology* 56 (6): 452–469.

Ariely, D., et al. 2000. The effects of averaging subjective probability estimates between and within judges. *Journal of Experimental Psychology. Applied* 6 (2): 130.

Aspinall, W. 2010. A route to more tractable expert advice. *Nature* 463: 264–265.

Batchelder, W. H., and R. Anders. 2012. Cultural consensus theory: Comparing different concepts of cultural truth. *Journal of Mathematical Psychology* 56 (5): 316–332.

Batchelder, W. H., and A. K. Romney. 1988. Test theory without an answer key. *Psychometrika* 53 (1): 71–92.

Bedford, T., and R. Cooke. R. 2001. *Probabilistic Risk Analysis: Foundations and Methods.* Cambridge University Press.

Beppu, A., and T. L. Griffiths. 2009. Iterated learning and the cultural ratchet. In *Proceedings of the 31st Annual Conference of the Cognitive Science Society.* Cognitive Science Society.

Bernstein, M. S., M. S. Ackerman, E. H. Chi, and R. C. Miller. 2011. The trouble with social computing systems research. In *CHI'11 Extended Abstracts on Human Factors in Computing Systems.* ACM.

Budescu, D., and E. Chen. 2014. Identifying expertise to extract the wisdom of crowds. *Management Science* 61 (2): 267–280.

Budescu, D. V., and A. K. Rantilla. 2000. Confidence in aggregation of expert opinions. *Acta Psychologica* 104 (3): 371–398.

Burgman, M. A., M. McBride, R. Ashton, A. Speirs-Bridge, L. Flander, B. Wintle, F. Fidler, L. Rumpff, and C. Twardy. 2011. Expert status and performance. *PLoS ONE* 6 (7): e22998.

Christensen-Szalanski, J., and J. B. Bushyhead. 1981. Physicians' use of probabilistic information in a real clinical setting. *Journal of Experimental Psychology: Human Perception and Performance* 7 (4): 928–935.

Cooke, R. M. 1991. *Experts in Uncertainty.* Oxford University Press.

Davis-Stober, C., D. Budescu, J. Dana, J., and S. Broomell. 2014. When is a crowd wise? *Decision* 1 (2): 79–101.

Dawid, A. P., and A. M. Skene. 1979. Maximum likelihood estimation of observer error-rates using the EM algorithm. *Applied Statistics* 28 (1): 20–28.

Ditta, A. S., and M. Steyvers. 2013. Collaborative memory in a serial combination procedure. *Memory* 21 (6): 668–674.

Einhorn, H. J. 1972. Expert measurement and mechanical combination. *Organizational Behavior and Human Performance* 7: 86–106.

Einhorn, H. J. 1974. Expert judgment: Some necessary conditions and an example. *Journal of Applied Psychology* 59: 562–571.

Engel, D., A. W. Woolley, L. X. Jing, C. F. Chabris, and T. W. Malone. 2014. Reading the mind in the eyes or reading between the lines? Theory of mind predicts collective intelligence equally well online and face-to-face. *PLoS ONE* 9 (12): e115212.

Fiser, J., P. Berkes, G. Orbán, and M. Lengyel. 2010. Statistically optimal perception and learning: From behavior to neural representations. *Trends in Cognitive Sciences* 14 (3): 119–130.

French, S. 1985. Group consensus probability distributions: A critical survey. In *Bayesian Statistics*, volume 2, ed. J. Bernardo, M. DeGroot, D. Lindley, and A. Smith. North-Holland.

French, S. 2011. Expert judgement, meta-analysis and participatory risk analysis. *Decision Analysis* 9 (2): 119–127.

Gallupe, R. B., L. M. Bastianutti, and W. H. Cooper. 1991. Unblocking brainstorms. *Journal of Applied Psychology* 76 (1): 137–142.

Gilovich, T. D. Griffin, and D. Kahneman, eds. 2002. *Heuristics and Biases: The Psychology of Intuitive Judgment*. Cambridge University Press.

Gigerenzer, G., U. Hoffrage, and H. Kleinbölting. 1991. Probabilistic mental models: A Brunswikian theory of confidence. *Psychological Review* 98: 506–528.

Goldstein, D. G., and G. Gigerenzer. 2002. Models of ecological rationality: The recognition heuristic. *Psychological Review* 109 (1): 75–90.

Griffiths, T. L., E. Vul, and A. N. Sanborn. 2012. Bridging levels of analysis for probabilistic models of cognition. *Current Directions in Psychological Science* 21 (4): 263–268.

Hertwig, R. 2012. Tapping into the wisdom of the crowd—with confidence. *Science* 336: 303–304.

Herzog, S. M., and R. Hertwig. 2009. The wisdom of many within one mind: Improving individual judgments with dialectical bootstrapping. *Psychological Science* 20: 231–237.

Hogarth, R. M. 1975. Cognitive processes and the assessment of subjective probability distributions. *Journal of the American Statistical Association* 70 (35): 271–289.

Hourihan, K. L., and A. S. Benjamin. 2010. Smaller is better (when sampling from the crowd within): Low memory span individuals benefit more from multiple opportunities for estimation. *Journal of Experimental Psychology: Learning, Memory, and Cognition* 36: 1068–1074.

Janis, I. L. 1972. *Victims of Groupthink: A Psychological Study of Foreign-Policy Decisions and Fiascoes*. Houghton Mifflin.

John, L. K., G. Loewenstein, and D. Prelec. 2012. Measuring the prevalence of questionable research practices with incentives for truth telling. *Psychological Science* 23 (5): 524–532.

Judd, S., M. Kearns, and Y. Vorobeychik. 2010. Behavioral dynamics and influence in networked coloring and consensus. *Proceedings of the National Academy of Sciences* 107 (34): 14978–14982.

Kahneman, D., P. Slovic, and A. Tversky. 1982. *Judgment under Uncertainty: Heuristics and Biases*. Cambridge University Press.

Kahneman, D., and A. Tversky, eds. 2000. *Choices, Values, and Frames*. Cambridge University Press.

Karger, D. R., S. Oh, and D. Shah. 2011. Iterative learning for reliable crowdsourcing systems. *Advances in Neural Information Processing Systems* 24: 1953–1961.

Kearns, M., S. Suri, and N. Montfort. 2006. An experimental study of the coloring problem on human subject networks. *Science* 313 (5788): 824–827.

Kearns, M., S. Judd, and Y. Vorobeychik. 2012. Behavioral experiments on a network formation game. In *Proceedings of the 13th ACM Conference on Electronic Commerce*. ACM.

Kerr, N. L., R. J. MacCoun, and G. P. Kramer. 1996. Bias in judgment: Comparing individuals and groups. *Psychological Review* 103 (4): 687–719.

Khatib, F., S. Cooper, M. D. Tyka, K. Xu, I. Makedon, Z. Popović, and F. Players. 2011. Algorithm discovery by protein folding game players. *Proceedings of the National Academy of Sciences* 108 (47): 18949–18953.

Koriat, A. 2012. When are two heads better than one and why? *Science* 336 (6079): 360–362.

Lee, M. D., E. Grothe, and M. Steyvers. 2009. Conjunction and disjunction fallacies in prediction markets. In *Proceedings of the 31th Annual Conference of the Cognitive Science Society*, ed. N. Taatgen, H. van Rijn, L. Schomaker, and J. Nerbonne. Erlbaum.

Lee, M. D., M. Steyvers, M. de Young, and B. J. Miller. 2012. Inferring expertise in knowledge and prediction ranking tasks. *Topics in Cognitive Science* 4: 151–163.

Lee, M. D., M. Steyvers, and B. Miller. 2014. A cognitive model for aggregating people's rankings. *PLoS ONE* 9 (5): e96431.

Lee, M. D., S. Zhang, and J. Shi. 2011. The wisdom of the crowd playing *The Price Is Right*. *Memory & Cognition* 39 (5): 914–923.

Libby, R., K. T. Trotman, and I. Zimmer. 1987. Member variation, recognition of expertise, and group performance. *Journal of Applied Psychology* 72 (1): 81–87.

Lichtendahl, K. C., Y. Grushka-Cockayne, and P. Pfeifer. 2013. The wisdom of competitive crowds. *Operations Research* 61 (6): 1383–1398.

Lin, S. W., and C. H. Cheng. 2009. The reliability of aggregated probability judgments obtained through Cooke's classical model. *Journal of Modelling in Management* 4 (2): 149–161.

Lipscomb, J., G. Parmigiani, and V. Hasselblad. 1998. Combining expert judgment by hierarchical modeling: An application to physician staffing. *Management Science* 44: 149–161.

Liu, Q., M. Steyvers, and A. Ihler. 2013. Scoring workers in crowdsourcing: How many control questions are enough? *Advances in Neural Information Processing Systems* 26: 1914–1922.

Lorenz, J., H. Rauhut, F. Schweitzer, and D. Helbing. 2011. How social influence can undermine the wisdom of crowd effect. *Proceedings of the National Academy of Sciences* 108 (22): 9020–9025.

Lorge, I., D. Fox, J. Davitz, and M. Brenner. 1958. A survey of studies contrasting the quality of group performance and individual performance, 1920–1957. *Psychological Bulletin* 55 (6): 337–372.

Mabe, P. A., and S. G. West. 1982. Validity of self-evaluation of ability: A review and meta-analysis. *Journal of Applied Psychology* 67 (3): 280–296.

Malone, T. W., R. Laubacher, and C. Dellarocas. 2010. The Collective Intelligence Genome. *Sloan Management Review* 51 (3): 21–31.

Mandel, D. R. 2005. Are risk assessments of a terrorist attack coherent? *Journal of Experimental Psychology: Applied* 11 (4): 277–288.

Mason, W. A., A. Jones, and R. L. Goldstone. 2008. Propagation of innovations in networked groups. *Journal of Experimental Psychology. General* 137 (3): 422–433.

Massey, C., J. P. Simmons, and D. A. Armor. 2011. Hope over experience: Desirability and the persistence of optimism. *Psychological Science* 22 (2): 274–281.

Mellers, B., L. Ungar, J. Baron, J. Ramos, B. Gurcay, K. Fincher, S. E. Scott, et al. 2014. Psychological strategies for winning a geopolitical forecasting tournament. *Psychological Science* 25 (5): 1106–1115.

Miller, B. J., and M. Steyvers. 2011. The wisdom of crowds with communication. In *Proceedings of the 33rd Annual Conference of the Cognitive Science Society*. Cognitive Science Society.

Miller, B. J., and M. Steyvers. 2014. Improving Group Accuracy Using Consistency across Repeated Judgments. Technical report, University of California, Irvine.

Norman, D. A. 1993. Distributed cognition. In *Things That Make Us Smart: Defending Human Attributes in the Age of the Machine*, ed. T. Dunaeff and D. Norman. Perseus Books.

Olson, K. C., and C. W. Karvetski. 2013. Improving expert judgment by coherence weighting. In *Proceedings of 2013 IEEE International Conference on Intelligence and Security Informatics*. IEEE.

Olson, G. M., T. W. Malone, and J. B. Smith, eds. 2001. *Coordination Theory and Collaboration Technology*. Erlbaum.

Ottaviani, M., and P. N. Sørensen. 2006. The strategy of professional forecasting. *Journal of Financial Economics* 81 (2): 441–466.

Prelec, D. 2004. A Bayesian truth serum for subjective data. *Science* 306: 462–466.

Prelec, D., and H. S. Seung. 2006. An algorithm that finds truth even if most people are wrong. Unpublished manuscript.

Rains, S. A. 2005. Leveling the organizational playing field—virtually: A meta-analysis of experimental research assessing the impact of group support system use on member influence behaviors. *Communication Research* 32 (2): 193–234.

Rauhut, H., and J. Lorenz. 2011. The wisdom of crowds in one mind: How individuals can simulate the knowledge of diverse societies to reach better decisions. *Journal of Mathematical Psychology* 55 (2): 191–197.

Ren, Y., and L. Argote. 2011. Transactive memory systems 1985–2010: An integrative framework of key dimensions, antecedents, and consequences. *Academy of Management Annals* 5 (1): 189–229.

Romney, A. K., W. H. Batchelder, and S. C. Weller. 1987. Recent applications of cultural consensus theory. *American Behavioral Scientist* 31 (2): 163–177.

Rowe, G., and G. Wright. 1999. The Delphi technique as a forecasting tool: Issues and analysis. *International Journal of Forecasting* 15 (4): 353–375.

Satopää, V. A., J. Baron, D. P. Foster, B. A. Mellers, P. E. Tetlock, and L. H. Ungar. 2014. Combining multiple probability predictions using a simple logit model. *International Journal of Forecasting* 30 (2): 344–356.

Shanteau, J., D. J. Weiss, R. P. Thomas, and J. C. Pounds. 2002. Performance-based assessment of expertise: How to decide if someone is an expert or not. *European Journal of Operational Research* 136 (2): 253–263.

Simmons, J. P., L. D. Nelson, J. Galak, and S. Frederick. 2011. Intuitive biases in choice versus estimation: Implications for the wisdom of crowds. *Journal of Consumer Research* 38 (1): 1–15.

Smyth, P., U. Fayyad, M. Burl, P. Perona, and P. Baldi. 1995. Inferring ground truth from subjective labeling of Venus images. *Advances in Neural Information Processing Systems*: 1085–1092.

Stankov, L., and J. D. Crawford. 1997. Self-confidence and performance on tests of cognitive abilities. *Intelligence* 25 (2): 93–109.

Stasser, G., and W. Titus. 1987. Effects of information load and percentage of shared information on the dissemination of unshared information during group discussion. *Journal of Personality and Social Psychology* 53 (1): 81–93.

Steiner, I. D. 1972. *Group Process and Productivity*. Academic Press.

Steyvers, M., M. D. Lee, B. Miller, and P. Hemmer. 2009. The wisdom of crowds in the recollection of order information. *Advances in Neural Information Processing Systems* 22: 1785–1793.

Steyvers, M., T. S. Wallsten, E. C. Merkle, and B. M. Turner. 2014. Evaluating probabilistic forecasts with Bayesian signal detection models. *Risk Analysis* 34 (3): 435–452.

Surowiecki, J. 2004. *The Wisdom of Crowds*. Random House.

Taleb, N. N. 2007. *The Black Swan: The Impact of the Highly Improbable*. Random House.

Tetlock, P. E. 2005. *Expert Political Judgment: How Good Is It? How Can We Know?* Princeton University Press.

Turner, B. M., M. Steyvers, E. C. Merkle, D. V. Budescu, and T. S. Wallsten. 2014. Forecast aggregation via recalibration. *Machine Learning* 95 (3): 261–289.

Tversky, A., and D. Kahneman. 1974. Judgment and uncertainty: Heuristics and biases. *Science* 185: 1124–1131.

Vul, E., and H. Pashler. 2008. Measuring the crowd within: Probabilistic representations within individuals. *Psychological Science* 19: 645–647.

Wallsten, T. S., and D. V. Budescu. 1983. State of the art—Encoding subjective probabilities: A psychological and psychometric review. *Management Science* 29 (2): 151–173.

Wallsten, T. S., D. V. Budescu, L. Erev, and A. Diederich. 1997. Evaluating and combining subjective probability estimates. *Journal of Behavioral Decision Making* 10: 243–268.

Wang, G., S. R. Kulkarni, H. V. Poor, and D. N. Osherson. 2011. Aggregating large sets of probabilistic forecasts by weighted coherent adjustment. *Decision Analysis* 8 (2): 128–144.

Watson, W. E., L. K. Michaelsen, and W. Sharp. 1991. Member competence, group interaction, and group decision making: A longitudinal study. *Journal of Applied Psychology* 76 (6): 803–809.

Weaver, R., and D. Prelec. 2013. Creating truth-telling incentives with the Bayesian Truth Serum. *Journal of Marketing Research* 50 (3): 289–302.

Wegner, D. M. 1987. Transactive memory: A contemporary analysis of the group mind. In *Theories of Group Behavior*, ed. B. Mullen and G. Goethals Springer.

Weiss, D. J., and J. Shanteau. 2003. Empirical assessment of expertise. *Human Factors* 45 (1): 104–116.

Weiss, D. J., K. Brennan, R. Thomas, A. Kirlik, and S. M. Miller. 2009. Criteria for performance evaluation. *Judgment and Decision Making* 4 (2): 164–174.

Welsh, M. B., M. D. Lee, and S. H. Begg. 2009. Repeated judgments in elicitation tasks: Efficacy of the MOLE method. In *Proceedings of the 31st Annual Conference of the Cognitive Science Society*. Cognitive Science Society.

Whitworth, B., B. Gallupe, and R. McQueen. 2001. Generating agreement in computer-mediated groups. *Small Group Research* 32 (5): 625–665.

Organizational Behavior

Editors' Introduction

Human organizations are among the most obvious examples of collective intelligence in the world around us. From small work teams to giant corporations, they have existed in some form for as long as humans have, and their influence on our lives today is hard to overstate. Because they coordinate the activities of many individual humans to achieve larger goals, organizations are among the easiest forms of collective intelligence to observe. Several disciplines study aspects of these organizations, including sociology, social psychology, economics, anthropology, communications, political science, and the interdisciplinary field—often called "organizational behavior"—that includes elements of all these other disciplines.

The study of human organizations is likely to be of particular interest to those who want to *design* new forms of collective intelligence because organization science, like computer science and parts of economics, is a "design science" (Simon 1969). In other words, unlike the study of intelligence as it occurs naturally in individual brains or animal groups, organization science is partly about how to *create* systems to achieve our goals.

The chapter that follows provides a comprehensive survey of what researchers have learned about how human organizations and teams are designed and how they work. For example, much of what researchers have learned about the structure and the processes of human organizations can apply to many other kinds of systems too. In designing any system, for instance, one has to decide both how to both *differentiate* the overall goals of the system into separable pieces and how to *integrate* the pieces by managing the interdependencies among them (Lawrence and Lorsch 1967; Thompson 1967).

But the design of human organizations is also strongly constrained by the properties of their most important building blocks: human beings. For example, understanding what motivates humans and creating appropriate incentives that appeal to these motivations is an important part of designing human organizations, but it may not generalize well to computer networks or to ant colonies. As another example, contrary to an assumption that is standard in all traditional economic models, monetary incentives can sometimes *decrease* a person's intrinsic motivations to do a task (see, e.g., Deci and Ryan 1985; Amabile 1993).

As is described in many of the other chapters in this volume, more and more important kinds of collective intelligence are becoming combinations of people and computers. Thus, it seems likely that insights from organizational researchers such as those described in this chapter will become increasingly useful in many of these other areas, too.

References

Amabile, T. M. 1993. Motivational synergy: Toward new conceptualizations of intrinsic and extrinsic motivation in the workplace. *Human Resource Management Review* 3 (3): 185–201.

Deci, E. L., and R. M. Ryan. 1985. *Intrinsic Motivation and Self-Determination in Human Behavior*. Plenum.

Lawrence, P. R., and J. W. Lorsch. 1967. *Organization and Environment : Managing Differentiation and Integration*. Harvard Business School Press.

Simon, Herbert A. 1969. *The Sciences of the Artificial*. MIT Press.

Thompson, J. D. 1967. *Organizations in Action*. McGraw-Hill.

Recommended Readings

Foundations

Donelson R. Forsyth. 2006. *Group Dynamics*. Thomson Wadsworth.

This widely used textbook provides a good overview of the factors that are important in understanding how human groups work, with an emphasis on small groups—such as work groups—containing fewer than about a dozen members. Some of these factors are unique to humans; others have clear analogs in other kinds of collectively intelligent groups.

Galbraith, J. R. 2002. *Designing Organizations*. Jossey-Bass.

This practitioner-oriented book provides a good overview of the factors that come into play when we move from designing small work groups to designing larger organizations that may have hundreds or thousands of members. As in Forsyth's book, some of the factors discussed (e.g., rewards) are specific to humans; others (e.g., structure) are much more general.

Recent Research

Kevin J. Boudreau, Nicola Lacetera, and Karim R. Lakhani. 2011. Incentives and problem uncertainty in innovation contests: An empirical analysis. *Management Science* 57 (5): 843–863.

One increasingly popular kind of "organization" today is found in online innovation contests. Unlike traditional hierarchies, these contests seek good ideas from large groups of people, far beyond the boundaries of a single legal organization. One important question in designing such contests is how many people to try to involve. Involving more people increases the chances that someone in the group will have a good idea; however, it also decreases the motivation for each participant, since each participant has less of a chance of winning. Boudreau, Lacetera, and Lakhani investigate this question empirically in the context of more than 9,000 software development contests and find that larger groups are desirable only when the problems being solved are highly uncertain.

Anita Williams Woolley, Christopher F. Chabris, Alexander Pentland, Nada Hashmi, and Thomas W. Malone. 2010. Evidence for a collective intelligence factor in the performance of human groups. *Science* 330 (6004): 686–688.

This paper illustrates how concepts from individual psychology (such as intelligence) can sometimes be directly applied to groups. It is well known in individual psychology that there is a single statistical factor, often called "general intelligence," that predicts how well a person will do on a wide variety of cognitive tasks. Wooley et al. ask whether there is a similar factor for groups. They find that there is, and they call it "collective intelligence." They also find that a group's collective intelligence is not strongly correlated with the individual intelligence of the members, but that it is correlated with the average social sensitivity of the members, with the equality in distribution of conversational turn-taking, and with the proportion of females in the group.

Collective Intelligence in Teams and Organizations

Anita Williams Woolley, Ishani Aggarwal, and Thomas W. Malone

In the 2014 Winter Olympic Games in Sochi, the Russian men's ice hockey team seemed poised to sweep their competition. With star players from the National Hockey League in North America and the Kontinental Hockey League in Russia, and with a home-field advantage in Russia, fans thought they were sure to win the gold medal. In fact, President Vladimir V. Putin declared that the success of the Sochi Olympics, which cost an estimated $50 billion, hinged on the success of the Russian men's hockey team. Not long into the tournament, however, it became clear that the team might not live up to the high expectations. Players who were high scorers on their professional teams didn't produce a single goal. Despite all its resources, talent, and drive, the team was eliminated from contention before the medal rounds even began. To make matters even worse, its final defeat was by the Finnish team, a previously undistinguished collection of third- and fourth-line professionals. Many people were dumbfounded: How could this team have failed so badly?

More than 30 years earlier, another hockey team from a different country had the opposite experience. In what came to be called the "Miracle on Ice," the 1980 U.S. men's hockey team, made up of amateurs and collegiate players, rose above all expectations and won the gold medal.

This distinction between talented individuals and talented teams is consistent with recent research documenting that a team's collective intelligence is a much stronger predictor of the team's performance than the ability of individual members (Woolley, Chabris, Pentland, Hashmi, and Malone 2010). Collective intelligence includes a group's capability to collaborate and coordinate effectively, and this is often much more important to the group's performance than individual ability alone. In other words, just having a number of smart individuals may be useful, but it is certainly not sufficient, for creating a smart group or a smart organization.

What are the necessary ingredients for collective intelligence to develop? In this chapter, we review frameworks and findings from the literature on team performance and the literature on organizational performance that may be especially useful to collective intelligence researchers for thinking about this question. To organize our review of the literature, we will use the Star Model of organizational design (Galbraith 2002). The Star Model identifies five categories of design choices that managers or other system designers can use to influence how an organization works:

Strategy—the overall goals and objectives the group or organization is trying to accomplish
Structure—how activities are grouped and who has decision-making power
Processes—the flow of information and activities among people, machines, and parts of the organization
Rewards—the motivation and incentives for individuals
People—the selection and development of the individuals and skills needed in the organization

Of course, the boundaries among these categories are somewhat arbitrary, and none of the design choices operate in isolation, so a successful organizational design depends, in part, on the proper alignment of all these elements. Our hope in this chapter is to give the reader a "tasting menu" of how these elements relate to one another, where to look for additional information about each, and how a system must have all five of these elements if it is to exhibit collective intelligence.

Strategy: Group Tasks and Goals

A group can be set up to fail by putting it to work on a task for which collective work isn't well suited or by giving the group unclear goals. The first step in designing any collectively intelligent system, therefore, is to make sure that the tasks the system is trying to perform or the goals it is trying to achieve are well suited to being worked on by a group (Locke, Durham, Poon, and Weldon 1997). Even when working on tasks well suited to group work, groups are often less than maximally effective because of *process loss*—that is, the additional difficulties they encounter because of sub-optimal processes (Steiner 1972).

Tasks that benefit from a variety of inputs and combined efforts tend to benefit from group collaboration. But simple tasks, and tasks that benefit from a high level of insight and coherence (such as many great works of art), are often better done by solo individuals.

Steiner (1972), as described by Forsyth (2006), identified four important types of group tasks on the basis of their structure:

conjunctive tasks, which operate at the level of the lowest performer (e.g., running in a group)
disjunctive tasks, which operate at level of the highest performer (e.g., answering math problems)
additive tasks, in which all contributions add to performance (e.g., shoveling snow)
compensatory tasks, in which, for instance, performance of one can offset mistakes of others (e.g., independent guesses to estimate a quantity such as the number of jelly beans in a jar).

Additive and compensatory tasks often benefit from groups working interdependently; disjunctive tasks can benefit from contributions of non-interacting groups, and highly skilled individuals are likely to outperform teams on conjunctive tasks. Furthermore, a task can be classified as *unitary* (meaning that it can't be divided into subtasks, and thus either the group must work on it all together or one person does the work while others watch) or *divisible* (meaning that the task can be efficiently or meaningfully divided into subtasks and assigned to group members) (Steiner 1972).

In addition, tasks can be characterized by the nature of the processes in which members of a group must engage in order to carry them out effectively (Larson 2009; McGrath 1984). For instance, McGrath's task circumplex identifies four task categories that exhibit different kinds of team interaction processes:

Generate tasks include creativity and planning tasks, both of which require idea generation. Usually, to succeed at generate tasks the members of a group should work in parallel to develop as many divergent ideas as they can.
Choose tasks or decision-making tasks require selecting among specified alternatives, either as in "intellective" tasks with an objectively correct answer, or as in "judgment" tasks with hard-to-demonstrate correct answers (Laughlin 1980; Laughlin and Ellis 1986). In either case, it is necessary for groups to engage in effective information sharing to

identify the correct response. (This will be discussed further in the section on processes below.)

Negotiate tasks involve resolving conflicts of interest or viewpoints.

Execute tasks involve performance of psychomotor tasks that require a high level of coordination, physical movement, or dexterity to produce a correct or optimal solution.

The type of task a group is faced with has important implications for the group's composition, incentives, structure, and process.

Regardless of the type of work to be accomplished by a team or an organization, two other important factors are the nature and the clarity of the goals being pursued. The positive effect of clear goals on individual performance is probably among the most-replicated results in all of organizational psychology (Locke and Latham 2006). Goals energize and direct behavior. Goal-directed people focus their attention on behaviors that are likely to lead to attainment of the goal and ignore activities that aren't relevant to the goal. Goals also arouse energy in proportion to their difficulty (up to the level of a worker's ability). The effects of goals are moderated by commitment—the effect on performance of a goal's difficulty increases with commitment to the goal. Goal specificity is also an important component; specific difficult goals produce better results than "do your best" or vague goals (Locke, Shaw, Saari, and Latham 1981).

The effects of clear goals are similarly strong at the group level (O'Leary-Kelly, Martocchio, and Frink 1994; Weldon and Weingart 1993), although the picture becomes somewhat more complex when one attempts to align individual goals with group goals. Whether group goals or individual goals are more salient, and whether they are aligned, determines the degree to which intragroup relations are characterized by cooperation or by competition. Other features of the task also come into play, such as whether the task is complex (Weingart 1992) and whether it requires members to work interdependently (Weldon and Weingart 1993). In addition, the types of goals assigned to groups have implications for the processes that develop. For instance, a team's strategic orientation (that is, whether its goals are more offensive or more defensive) has implications for the kinds of information that will be attended to within the group or in the environment (Woolley 2011).

The Structure of a Group or an Organization

Coordination is one of the most important problems a group or an organization must solve in order to be effective (March and Simon 1958). The need to coordinate members' activities arises from the interdependent nature of the activities that the members perform (Argote 1982). Faraj and Xiao (2006) concluded that "at its core, coordination is about the integration of organizational work under conditions of task interdependence and uncertainty" (see also Okhuysen and Bechky 2009). More specifically, in teams the word 'coordination' often refers to the process of synchronizing or aligning members' activities with respect to their sequence and timing (Marks, Mathieu, and Zaccaro 2001; Wittenbaum, Vaughan, and Stasser 1998).

Furthermore, though groups and organizations can coordinate through the explicit development of plans and routines, dynamic situations often call for planning that occurs in real time (Wittenbaum et al. 1998). For instance, when discussing medical emergency units Argote (1982) argued that non-programmed means of coordination, which involve on-the-spot sharing of information among members, are effective in dealing with the increased demands associated with increased uncertainty.

Though the ability to coordinate tacitly and dynamically may be an important contributor to collective intelligence, it may also be an outcome. In a study of tacit coordination in laboratory teams, Aggarwal, Woolley, Chabris, and Malone (2015) found that collective intelligence was a significant predictor of teams' ability to coordinate their choices in a behavioral economics game, despite being unable to communicate, allowing some groups to earn significantly more money than others during the lab session.

One of the most important ways in which members of groups and organizations coordinate is through their structure. And the larger the group, the more important structure can be in determining the group's effectiveness.

At a very high level, organizational theorists and economists have made a distinction between organizing activities in *hierarchies* and in *markets* (Williamson 1981). For instance, the activity of producing automobile tires can, in principle, be performed inside the same hierarchical organization (say, General Motors) that manages other parts of the process of producing cars, or it can be performed by an external supplier (say, Goodyear). In the former case, the activity is coordinated by hierarchical management processes inside the firm (General Motors); in the latter, it is coordinated by negotiations in a market and contracts between a buyer (General Motors) and a seller (Goodyear).

The choice of an arrangement depends crucially on the *transaction costs* of the different arrangements, and these costs, in turn, are affected by opportunism, search costs, and the specificity of the assets exchanged

(Williamson 1973). Some authors have also talked about other kinds of organizational structures, such as *networks* in which rapidly shifting connections within a single organization or among different organizations are much more important than the stable hierarchies of traditional organizations (Powell 1990).

The vast majority of research on organizational structure has focused on ways of structuring hierarchical organizations. Several lessons about collective intelligence in general emerge from this work:

Differentiation and integration are both needed. As Lawrence and Lorsch (1967) point out, effective organizations usually have to differentiate— that is, to divide the overall goal of the organization into different kinds of tasks and to create different parts of the organization that are focused on these different kinds of work. For instance, the division of labor might involve creating different groups for marketing, manufacturing, and engineering, or for different products or customers. But then there also has to be some way to *integrate* the different parts of the organization to achieve the organization's overall goals (ibid.). For instance, organizations can coordinate the activities of different organizational parts using mechanisms such as *informal lateral communication* (such as casual conversations at lunch), *formal groups* (such as task forces), *integrating managers* (such as product managers or account managers), or *matrix managers* (Galbraith 2002).

Integration can be viewed as managing interdependencies. Thompson (1967) identified three types of interdependencies among activities: *pooled* interdependencies (in which activities share a resource, such as money or machine time), *sequential* interdependencies (in which resources from one activity flow to another one), and *reciprocal* interdependencies (in which resources flow back and forth between two or more activities). Thompson and later researchers—among them Malone et al. (1999) and Van de Ven et al. (1976)—showed how different kinds of coordination processes are appropriate for different kinds of interdependencies. For instance, pooled (or "shared resource") dependencies can be managed by coordination processes such as "first come-first served," priority order, budgets, managerial decision, or market-like bidding (Malone et al. 1999).

There is no one best way to organize. The widely accepted contingency theory of organization design (Lawrence and Lorsch 1967; Thompson 1967; Galbraith 1973) holds that there is no one best way to organize. Instead, according to this view, the best organizational design for a given situation depends on many factors, including the organization's strategy, tasks, technology, customers, labor markets, and other aspects of its environment (see, e.g., Daft 2001; Duncan 1979). For instance, *functional structures* (with separate departments for functions like engineering, manufacturing, and sales) are well suited to situations where maximizing depth of functional expertise and economies of scale are critical, but they are generally not well suited to situations where rapid adaptation to changing environments is important. *Divisional structures* (with separate divisions for different products, customers, or geographical regions) are well suited to environments where rapid adaptation to environmental changes is important or where success depends on customizing products or services for specific types of customers or regions. But they are not well suited to reducing costs by taking advantage of economies of scale. In *matrix structures*, there are both functional and divisional structures, and some employees report to two (or more) bosses. For instance, an engineering manager might report to both a vice-president of engineering and a vice-president for a specific product. The matrix structure has the potential to achieve the benefits of both functional and divisional structures (such as both economies of scale and rapid adaptation to change), but it involves significantly more managerial complexity and coordination costs.

Although these principles of organizational design were articulated in the context of large, hierarchical human organizations, we suspect that they can all be generalized in ways that could help us to understand collective intelligence in many other kinds of systems, such as computer networks, brains, and ant colonies.

In addition to analyzing traditional, hierarchical organizations, some organizational researchers have also begun to analyze the new kinds of organizational forms that are beginning to emerge as new information technologies make possible new ways of organizing human activity (Malone 2004; Malone, Yates, and Benjamin 1987). For example, more decentralized structures such as loose hierarchies, democracies, and markets may become more common as inexpensive communication technologies make them more feasible (Malone 2004). Malone, Laubacher, and Dellarocas (2010) have identified a set of design patterns (or "genes") that arise repeatedly in innovative new forms of collective intelligence, such as Wikipedia, InnoCentive, and open-source software communities, such as Linux. Examples of these "genes" include contests, collaborations, prediction markets, and voting.

Processes

The aspects of group process most germane to collective intelligence are those that characterize intelligent systems more generally, whether technological or biological: memory, attention, problem solving, and learning. Analogous processes in each of these categories have been explored at the group level. This is consistent with the emerging view of groups as information processors (Hinsz, Tindale, and Vollrath 1997) in that many of the group processes most central to group functioning involve cognitive or meta-cognitive processes.

Memory in Groups

Memory in groups has been studied mainly via work on transactive memory systems. The term "transactive memory system" (TMS) refers to a shared system that individuals in groups develop to collectively encode, store, and retrieve information or knowledge in different domains (Wegner 1987; Argote and Ren 2012; Hollingshead 2001; Lewis and Herndon 2011). A group with a well-developed TMS can efficiently store and make use of a broader range of knowledge than groups without a TMS. According to TMS theory as conceived by Wegner (1987), and first demonstrated in the context of small groups by Liang, Moreland, and Argote (1995), there are three behavioral indicators of TMS: specialization, credibility, and coordination.

Specialization in the team is characterized by how group members divide the cognitive labor of their tasks, with different members specializing in specific domains. Credibility is characterized by members' reliance on one another to be responsible for specific expertise such that collectively they possess all of the information needed for their tasks. Coordination is characterized by smooth and efficient action (Lewis 2004; Moreland, Argote, and Krishnan 2002; Moreland and Myaskovsky 2000).

Through performing tasks and answering questions, a member establishes credibility and expertise status. Other members, aware of that member's expertise, direct new knowledge in the domain to him or her, which reinforces his or her specialization and team members' trust in his or her expertise. Further, members know whom to count on for performing various tasks and whom to consult for information in particular domains, which improves coordination (Argote and Ren 2012). Dozens of studies have demonstrated the positive effects of TMS on group performance in both laboratory and field settings (Lewis and

Herndon 2011), though work continues to refine measures and conceptualization of the construct and its relationship to performance for different types of tasks (ibid.).

Attention in Groups

In individuals, teams, and organizations, attention is viewed as central to explaining the existence of a limited information-processing capacity (Ocasio 2011; Styles 2006) and thus has a great deal of relevance to understanding and studying collective intelligence. Work on attention at the organizational level started with the work of Simon (1947), who examined the channeling, structuring, and allocation of attention in administrative behavior. Cohen, March, and Olsen (1972) examined attention allocation in organizational decision making. Ocasio (1997) focused on how attention in organizations shapes organizational adaptation.

In a more recent review of the developing literature on attention in organizations, Ocasio (2011) identified three different theoretical lenses that have been used in studying attention:

attentional perspective (i.e., the top-down cognitive structures that generate heightened awareness and focus over time to relevant stimuli and responses)
attentional engagement (i.e., sustained allocation of cognitive resources to guide planning, problem solving, sense making, and decision making)
attentional selection (i.e., the emergent outcome of processes that result in focusing attention on selected stimuli or responses to the exclusion of others).

Newer lines of work examine the development of shared attention in groups under the "attentional selection" category identified by Ocasio (2011), and ask: What do teams make the center of their focus as they conduct their work? What do they allow to fall by the wayside?

Teams exhibit regularities in the types of issues they attend to in the course of their work, and these regularities have been studied as part of research on task focus in teams. Some teams are *process-focused*, focusing on the specific steps necessary to carry out tasks and how those are arranged among members and over time (Woolley 2009a,b). *Outcome-focused* teams place more emphasis on the products of their work or the "big picture" and allow that to drive coordination and decision making.

Teams that are high in outcome focus tend to produce more innovative or creative outcomes, and adapt more effectively to difficulties that arise in their work (Woolley 2009a); process-focused teams commit fewer errors (Aggarwal and Woolley 2013). More recent work on offensive and defensive strategic orientation shows that a team's position in a competitive environment is an important contextual antecedent of outcome or process focus and the balance of attention members pay to the internal workings of the group versus the environment (Woolley 2011; Woolley, Bear, Chang, and DeCostanza 2013).

Not only is the content of team focus important; so is the degree to which members agree about it. This agreement around strategic priorities has been called "team strategic consensus" in laboratory teams (Aggarwal and Woolley 2013) and "strategic consensus" in top management teams (Floyd and Wooldridge 1992; Kellermanns, Walter, Lechner, and Floyd 2005); at the dyadic level, it has been called "strategic compatibility" (Bohns and Higgins 2011). The degree to which members of a group agree about the group's strategic priorities is likely to affect the clarity with which they will execute the task. Agreement on process focus, for example, has been shown to be extremely beneficial to reducing errors in production tasks and in execution tasks (Aggarwal and Woolley 2013) but to undermine the development of creative outcomes in teams (Aggarwal and Woolley 2012).

Problem Solving and Decision Making by Groups

Comprehensive treatments of problem solving by groups encompass much of what we have already discussed in this chapter, including group goals, task types, and social processes (Laughlin 1980). And the view of groups as entities that process information and make decisions is increasingly central to research on group problem solving (Hinsz, Tindale, and Vollrath 1997). The ability of groups to process information effectively—that is, to share relevant details, weight information appropriately, and arrive at the best conclusion—is directly tied to team performance (Mesmer-Magnus and DeChurch 2009). Groups often base their decisions on irrelevant information and disregard relevant information (Larson 2009). Thus factors affecting the quality of group decision making have direct implications for collective intelligence.

The main problems experienced in group decision making are associated with surfacing the relevant information and combining it appropriately. Surfacing the relevant information is complicated by many of the issues concerned with other aspects of the Star Model: Is there enough diversity of group members to have access to all of the necessary information? Are the members' goals and motivations aligned enough that they are willing to share the information they have? If these questions have been suitably addressed, there are a range of cognitive, motivational, and affective factors that can influence the kinds of information groups attend to (or ignore) in decision making. In terms of cognitive factors, a long line of work on social decision schemes has investigated how pre-decision preferences of individuals combine to influence a joint decision (Davis 1973). Groups are also more likely than individual decision makers to use certain cognitive heuristics and biases (Kerr, MacCoun, and Kramer 1996). In particular, groups are vulnerable to biases resulting from the initial distribution of information. For instance, when there are "hidden profiles," in which members initially prefer different alternatives on the basis of conflicting information they hold, they may have to make a special effort to surface and share all the information they need to reach the correct solution (Stasser and Titus 1985).

Motivational approaches to group decision making focus on group members' motivation to overlook disconfirming evidence and to believe in the infallibility of their group. Isenberg (1986), Janis and Mann (1977), and Sanders and Baron (1977) have examined these motivational issues in their work on groupthink and social comparison. Toma and Butera (2009) have demonstrated that within-group competition leads group members to share less information, and to be less willing to disconfirm initial preferences, as a result of mistrusting their teammates.

Combining social and motivational factors, De Dreu et al. (2008) proposed a theory of motivated information processing in groups, in which epistemic motivation (motivation to understand the world) determines how deeply group members seek out information, and social motivations (such as cooperation and competition) determine what information is shared by the group. Thus, epistemic and social motivations interact to affect the quality of group judgment and group decision making.

In view of the biases and difficulties in group decision making, Surowiecki (2004) and others have advocated using collections of *independent* decision makers to gain the advantage of multiple perspectives without the drawbacks of the social processes that bias decisions. First demonstrated by Galton (1907), it has since been repeatedly shown in studies of guessing and problem solving that the average of many individuals' estimates is often closer to the true value than almost all of the individual or even expert guesses. However, for any benefits to accrue from the use of a crowd, the individual estimates must be independent of one another and the sample sufficiently large and unbiased to enable errors to be symmetrically distributed (Surowiecki 2004). Even subtle social influence revealing knowledge of others' estimates can create a cascade of effects that reduces the accuracy of crowds (Lorenz, Rauhut, Schweitzer, and Helbing 2011).

Although independent decision makers can be useful for some types of decisions when the conditions for accuracy are in place, there are other circumstances in which traditional interacting groups are usually better at making decisions. For instance, interacting groups are often better when the options are not well defined or when the group needs to buy into a decision for it to be implemented. In these circumstances, a number of interventions have been demonstrated to successfully improve group decision making. One type of intervention focuses on structuring group conversation so that the group identifies questions and how their information must be integrated to answer those questions (see, e.g., Woolley, Gerbasi, Chabris, Kosslyn, and Hackman 2008). This approach can also be operationalized in the form of decision-support systems that structure members' inputs and facilitate integration.

A second type of intervention in group decision making involves putting group members into different roles to adopt opposing points of view. These are known most generally as "devil's advocate" approaches. They were named after a similar process adopted during the sixteenth century as part of the canonization process in the Roman Catholic Church. In the canonization process, an appointed person (the devil's advocate) would take a skeptical view of a candidate in opposition to God's advocate, who argued in the candidate's favor.

A third approach involves encouraging a group to grant equal speaking time to all members on the assumption that this will enable them to bring more relevant facts into the discussion. Equality in speaking time has been associated with higher collective intelligence in groups (Engel et al. 2014; Woolley et al. 2010). Interventions involving real-time feedback on relative contributions to group conversation have also been shown to improve a group's decision-making performance (DiMicco et al. 2004).

Group Learning
Some views of intelligence equate the concepts of intelligence and learning. For instance, in individual psychology the information-processing viewpoint on intelligence sees learning as a basic process of intelligence (Sternberg and Salter 1982). Similarly, research on

organizational IQ operationalizes the measure as the ability of the organization to gain new knowledge from R&D investments (Knott 2008). However, other work conceptualizes learning as one outcome of the basic capability of collective intelligence (Aggarwal, Woolley, Chabris, and Malone 2015).

Whether learning is encompassed within intelligence or viewed as an outcome of it, a great deal of evidence suggests that groups and organizations vary enormously in their ability to learn. With experience, the performance of some organizations improves dramatically but that of other organizations remains unchanged or even deteriorates (Argote 1999).

In general, the term "group learning" refers to changes in a group—including changes in cognitions, routines, or performance—that occur as a function of experience (Argote, Gruenfeld, and Naquin 2001; Argote and Miron-Spektor 2011; Fiol and Lyles 1985). For example, as a group gains experience, it may acquire information about which of its members are good at certain tasks, how to use a new piece of technology more effectively, or how to coordinate their activities better. This knowledge may, in turn, improve the group's performance (Argote 1999).

It is sometimes useful to distinguish two kinds of group learning: changes in *knowledge* (which may be gauged from change in performance) and changes in *group processes* or repertoires (Argote et al. 2001; Argote and Miron-Spektor 2011; Edmondson 1999; Fiol and Lyles 1985; Wilson et al. 2007). It is also important to recognize that a group may learn (e.g., change processes) without any change in its performance, and that a group's performance may change (perhaps because of changes in the environment) without any corresponding change in the group's knowledge (Argote 1999). And though sometimes knowledge may be explicit (easily codifiable and observable; see, e.g., Kogut and Zander 1992), at other times it may be only tacit (unarticulated and difficult to communicate; see, e.g., Nonaka 1994).

An organization's overall ability to learn productively—that is, to improve its outcomes through better knowledge and insight (Fiol and Lyles 1985)—depends on the ability of the smaller groups within it to learn (Edmondson 1999; Roloff, Woolley, and Edmondson 2011; Senge and Sterman 1992). Much of the work on group learning uses a concept of learning curves that was originally developed in individual psychology (Ebbinghaus 1885; Thorndike 1898) to characterize the rate of improvement, and researchers have found considerable variation in that rate among groups (Argote and Epple 1990; Dutton and Thomas 1984; Knott 2008).

Motivation and Incentives

If a group is working on a well-defined task, is structured appropriately, and is using effective processes for conducting work, it is also important to evaluate whether the group's members are properly motivated to do the work. As we discussed above, specific difficult goals can be motivating, but motivation also can come from other sources. The literature has generally looked at two sources of motivation: extrinsic motivation, which is often in the form of monetary incentives, and intrinsic motivation, which is derived from the internal satisfaction associated with the work itself.

Monetary incentives are one of the most common ways to induce high levels of effort in traditional organizational settings (Lazear 2000; Prendergast 1999). At times they have been shown to increase the quantity but not the quality of work (see, e.g., Jenkins, Mitra, Gupta, and Shaw 1998). The use of group-level monetary incentives can be tricky, as group-based incentives are highly subject to free riding (Alchian and Demsetz 1972; Lazear and Shaw 2007). Creating reward interdependence in teams can enhance performance, but only if the work behavior is highly cooperative (Wageman 1995; Wageman and Baker 1997).

When it is difficult for an employer to identify and reward each employee's exact contribution to the team's output, employees working in a team typically will lack incentives to provide the optimal level of effort and will work less than if they were working alone. This suggests that collaboration, particularly by anonymous workers outside of long-term relationship, may produce moral hazard (Holmstrom 1982) and social loafing (Latane, Williams, and Harkins 1979).

The potential for moral hazard may be exacerbated in the group work typical of online platforms, which could attract individuals with any number of characteristics and inclinations—including those more inclined toward free riding (Kerr and Bruun 1983). But despite the risk of free riding, monetary incentives can often be effective, even when they are based on group—not individual—outcomes, and especially when compared to incentive schemes that don't depend on outcomes at all (Prendergast 1999).

The research relating incentives to the creativity of teams is a bit muddled. Whereas some studies (see, e.g., Eisenberger and Rhoades 2001) suggest that extrinsic or cash rewards for teams promote creativity, others (see, e.g., Kruglanski, Friedman, and Zeevi 1971; Manso

2011) suggest that extrinsic rewards inhibit creativity or produce other undesirable effects.

Cash incentives can also, at times, crowd out non-monetary motivations (see, e.g., Frey and Jegen 2001), which are especially important in the case of creative problem-solving work. Teresa Amabile and colleagues have demonstrated that reduced intrinsic motivation and reduced creativity can be caused by any of several extrinsic factors, including expected external evaluation, competition with peers, and constrained choice in how to do one's work. Though competing with peers (who might otherwise share information) seems to dampen creativity, competing with outside groups or organizations can stimulate it (see Amabile and Fisher 2000 for a review).

There are also circumstances in which certain forms of extrinsic motivation may support intrinsic motivation and creativity—or at least not undermine it (Amabile 1993). This "motivational synergy" is most likely to occur when people feel that the reward confirms their competence and the value of their work, or enables them to do work they were already interested in doing. This is consistent with earlier research demonstrating that "informational" and "enabling" rewards can have positive effects on intrinsic motivation (Deci and Ryan 1985).

It is also important to recognize that monetary incentives do not preclude other motivations. For instance, in peer production (e.g., development of open-source software) there are many conspicuous non-monetary motivations for participants, including the intrinsic enjoyment of doing the task, benefits to the contributors from the using the software or other innovations themselves, and "social" motivations fed by the presence of other participants on the platform (Lakhani and Wolf 2005). "Social" motivations include such things as an interest in gaining affiliation with the larger team as a community and an interest in accruing status or signaling one's expertise to the community (Butler, Sproull, Kiesler, and Kraut 2007; Lakhani and Wolf 2005; Lerner and Tirole 2005).

Evidence also suggests that online collaboration, rather than necessarily attracting loafers, may attract people who prefer to collaborate and thus will work relatively diligently (Boudreau, Lacetera, and Lakhani 2011). In fact, online collaboration often embodies the job characteristics that Hackman and Oldham (1976) found to be most directly associated with internal motivation: variety of content, autonomy in how work is conducted, and knowledge of results.

Selecting the Right People

We now come to the last component of the Star Model: the selection of the right individuals to carry out the work. Two categories of characteristics are important to consider when selecting members of a team or an organization with an eye toward enhancing collective intelligence: characteristics that contribute information or skills to the group (and thus must be considered in combination with other members) and characteristics that facilitate the transfer of information (and can be evaluated individually).

A long line of research on group diversity has examined the types of differences that are helpful to group performance and the types of differences that are harmful. The information-processing perspective suggests that a diverse team with a relatively broad range of task-relevant knowledge, skills, and abilities has a larger pool of resources for dealing with non-routine problems (Van Knippenberg and Schippers 2007; Williams and O'Reilly 1998). In fact, one of the primary reasons organizations use teams, and not simply individuals, is to have access to a diverse array of information, perspectives, and skills. For that reason, group composition is one of the most commonly studied team variables (see, e.g., Guzzo and Dickson 1996; Hollenbeck, DeRue, and Guzzo 2004; Reiter-Palmon, Wigert, and Vreede 2012; Tesluk, Farr, and Klein 1997).

Despite diversity's potential value, however, a number of studies and meta-analyses (e.g., Joshi and Roh 2009) have failed to show that it has a strong effect on a team's performance. Scholars have, therefore, urged researchers to pay close attention to the type of diversity variable they study. It may be critical, for example, to examine the specific type of diversity that is most relevant to the outcomes being investigated (Harrison and Klein 2007; Horwitz and Horwitz 2007; Joshi and Roh 2009; Milliken and Martins 1996).

A group performing a task that benefits from a range of skills or expertise will underperform unless composed with the requisite cognitive diversity (Woolley et al. 2007, 2008), even in comparison with groups of higher general intelligence or ability (Hong and Page 2004). Groups that are too homogenous will also be less creative than more cognitively diverse groups (Aggarwal and Woolley 2012) and will exhibit lower levels of collective intelligence than moderately cognitively diverse groups (Aggarwal et al. 2015). However, cognitively diverse groups run the risk of making errors in execution, particularly when the diversity leads them to disagree as to how to prioritize elements of tasks (Aggarwal and Woolley 2013). Thus, many research-ers focus on the moderating effects of group processes, such as the development of transactive memory systems and strategic consensus, when examining the relationship between diversity and performance.

Other important characteristics to consider when assembling a group are related to social or emotional intelligence. Emotional intel-ligence is defined as the capacity to reason about emotions and to use emotions to enhance thinking. It includes the abilities to accurately perceive emotions in others, to access and generate emotions so as to assist thought, to understand emotions and emotional knowledge, and to reflectively regulate emotions so as to promote emotional and intel-lectual growth (Mayer and Salovey 1993). There is a general consensus that emotional intelligence enhances a group's performance (Druskat and Wolff 2001), at least in the short term (Ashkanasy and Daus 2005).

A specific subset of skills related to the perception of emotions and mental states has been studied under the rubric "theory of mind" (ToM) (Apperly 2012; Baron-Cohen et al 2001; Flavell 1999; Premack and Woodruff 1978; Saxe 2009). Theory-of-mind ability encompasses the accurate representation and processing of information about the mental states of other people, also known as "mentalizing ability" (Baron-Cohen et al. 2001), which contribute to successful interaction. Therefore, theory of mind appears to be the component of emotional intelligence most relevant to studies of collective intelligence.

The ability to make simple inferences about the false beliefs of others has been explored by developmental psychologists as a milestone reached by pre-school-age children (Wimmer and Perner 1983), and it is widely recognized that people with autism and various other clinical conditions have difficulties with theory of mind (Baron-Cohen 1991). A common (though usually untested) assumption in much of this research is that people with greater theory-of-mind abilities will be more competent at various kinds of social interaction. But only a few studies have tested this in limited ways with children (Begeer et al. 2010; Peterson et al. 2007; Watson et al. 1999), and fewer still have tested it with adults (Bender et al. 2012; Krych-Appelbaum et al. 2007; Woolley et al. 2010).

For instance, Woolley et al. (2010) found that groups whose members had higher average ToM scores as measured by the "Reading the Mind in the Eyes" (RME) test (Baron-Cohen et al. 2001) also had significantly higher collective intelligence. Indeed, average ToM scores remained the only significant predictor of collective intelligence even when individual intelligence or other group composition or process variables, such as proportion of women in the group or distribution of communication, was controlled for.

The degree to which theory-of-mind abilities, as measured by RME or otherwise, can be altered by training or experience remains an open question. Recent studies (see, e.g., Kidd and Castano 2013) suggest that theory-of-mind abilities as measured by the RME test can be, at least temporarily, improved by reading literary fiction, which implies a new and interesting avenue of research for improving group performance.

Conclusion

In this chapter, we have provided a brief and selective overview of a relatively vast literature on group and organizational performance. We have focused specifically on variables that strike us as particularly germane for the design and study of collectively intelligent systems. In so doing, we have used Galbraith's Star Model to guide our consideration of the various issues to be considered by effective organizations.

It is intriguing to further consider how human systems or human–computer systems might deal with these issues in completely new ways. For instance, could we design human–computer environments in such a way that group processes would be automatically structured to be optimal for the type of task facing a group at a given time? So that developing transactive memory systems in groups would be either automatic or trivial? So that group members would be prompted to balance their contributions to the work at hand and matched perfectly in terms of their distribution of knowledge or skills? So that subtle social cues would be amplified in a manner to allow the group as a whole to enjoy a high level of emotional intelligence? These are only a few of the possibilities that are suggested by coupling an understanding of the key factors for collective intelligence identified in the literature on teams and organizations with those of other literatures discussed in this volume. We hope the research and the ideas discussed here will help readers to imagine new ways of increasing collective intelligence that go far beyond anything that was possible before.

References

Aggarwal, I., and A. W. Woolley. 2012. Cognitive style diversity and team creativity. Presented at seventh annual INGRoup conference, Chicago.

Aggarwal, I., and A. W. Woolley. 2013. Do you see what I see? The effect of members' cognitive styles on team processes and performance. *Organizational Behavior and Human Decision Processes* 122: 92–99.

Aggarwal, I., A. W. Woolley, C. F. Chabris, and T. W. Malone. 2015. Cognitive diversity, collective intelligence, and learning in teams. In Proceedings of Collective Intelligence 2015, Santa Clara.

Alchian, A. A., and H. Demsetz. 1972. Production, information costs, and economic organization. *American Economic Review* 62 (5): 777–795.

Amabile, T. M. 1993. Motivational synergy: Toward new conceptualizations of intrinsic and extrinsic motivation in the workplace. *Human Resource Management Review* 3 (3): 185–201.

Amabile, T. M., and C. M. Fisher. 2000. Stimulate creativity by fueling passion. In *Handbook of Principles of Organizational Behavior*, second edition, ed. E. Locke. Wiley.

Apperly, I. A. 2012. What is "theory of mind"? Concepts, cognitive processes and individual differences. *Quarterly Journal of Experimental Psychology* 65 (5): 825–839.

Argote, L. 1982. Input uncertainty and organizational coordination in hospital emergency units. *Administrative Science Quarterly* 27 (3): 420–434.

Argote, L. 1999. *Organizational Learning: Creating, Retaining, and Transferring Knowledge.* Kluwer.

Argote, L., and D. Epple, D. 1990. Learning curves in manufacturing. *Science* 247: 920–924.

Argote, L., D. Gruenfeld, and C. Naquin. 2001. Group learning in organizations. In *Groups at Work: Advances in Theory and Research*, ed. M. E. Turner. Erlbaum.

Argote, L., and E. Miron-Spektor. 2011. Organizational learning: From experience to knowledge. *Organization Science* 22 (5): 1123–1137.

Argote, L., and Y. Ren. 2012. Transactive memory systems: A microfoundation of dynamic capabilities. *Journal of Management Studies* 49 (8): 1375–1382.

Ashkanasy, N. M., and C. S. Daus. 2005. Rumors of the death of emotional intelligence in organizational behavior are vastly exaggerated. *Journal of Organizational Behavior* 26 (4): 441–452.

Baron-Cohen, S. 1991. The theory of mind deficit in autism: How specific is it?*. *British Journal of Developmental Psychology* 9 (2): 301–314.

Baron-Cohen, S., S. Wheelwright, J. Hill, Y. Raste, and I. Plumb. 2001. The "Reading the Mind in the Eyes Test" revised version: A study with normal adults, and adults with Asperger syndrome or high-functioning autism. *Journal of Child Psychology and Psychiatry* 42 (02): 241–251.

Begeer, S., B. F. Malle, M. S. Nieuwland, and B. Keysar. 2010. Using theory of mind to represent and take part in social interactions: Comparing individuals with high-functioning autism and typically developing controls. *European Journal of Developmental Psychology* 7 (1): 104–122.

Bender, L., G. Walia, K. Kambhampaty, K. E. Nygard, and T. E. Nygard. 2012. Social sensitivity and classroom team projects: An empirical investigation. In *Proceedings of the 43rd ACM Technical Symposium on Computer Science Education.* ACM.

Bohns, V. K., and E. T. Higgins. 2011. Liking the same things, but doing things differently: Outcome versus strategic compatibility in partner preferences for joint tasks. *Social Cognition* 29 (5): 497–527.

Boudreau, K. J., N. Lacetera, and K. R. Lakhani. 2011. Incentives and problem uncertainty in innovation contests: An empirical analysis. *Management Science* 57 (5): 843–863.

Butler, B., L. Sproull, S. Kiesler, and R. Kraut. 2007. Community building in online communities: Who does the work and why? In *Leadership at a distance*, ed. S. Weisband. Erlbaum.

Cohen, M. D., J. G. March, and J. P. Olsen. 1972. A garbage can model of organizational choice. *Administrative Science Quarterly* 17 (1): 1–25.

Daft, R. L. 2001. *Essentials of Organization Theory and Design*. South-Western.

Davis, J. H. 1973. Group decision and social interaction: A theory of social decision schemes. *Psychological Review* 80 (2): 97–125.

Deci, E. L., and R. M. Ryan. 1985. *Intrinsic Motivation and Self-Determination in Human Behavior*. Plenum.

De Dreu, C. K., B. A. Nijstad, and D. van Knippenberg. 2008. Motivated information processing in group judgment and decision making. *Personality and Social Psychology Review* 12 (1): 22–49.

DiMicco, J. M., A. Pandolfo, and W. Bender. 2004. Influencing group participation with a shared display. In *Proceedings of the 2004 ACM Conference on Computer Supported Cooperative Work*. ACM.

Druskat, V. U., and S. B. Wolff. 2001. Building the emotional intelligence of groups. *Harvard Business Review* 79 (3): 80–90.

Duncan, R. B. 1979. What is the right organizational structure? Decision tree analysis provides the answer. *Organizational Dynamics*, winter: 59–79.

Dutton, J. M., and A. Thomas. 1984. Treating progress functions as a managerial opportunity. *Academy of Management Review* 9: 235–247.

Ebbinghaus, H. 1885. *Memory: A Contribution to Experimental Psychology*. Dover.

Edmondson, A. 1999. Psychological safety and learning behavior in work teams. *Administrative Science Quarterly* 44 (2): 350–383.

Eisenberger, R., and L. Rhoades. 2001. Incremental effects of reward on creativity. *Journal of Personality and Social Psychology* 81 (4): 728–741.

Engel, D., A. W. Woolley, L. Jing, C. F. Chabris, and T. W. Malone. 2014. Reading the mind in the eyes or reading between the lines? Theory of mind predicts collective intelligence equally well online and face-to-face. *PLoS ONE* 9 (12): e115212.

Faraj, S., and Y. Xiao. 2006. Coordination in fast-response organizations. *Management Science* 52 (8): 1155–1169.

Fiol, C. M., and M. A. Lyles. 1985. Organizational learning. *Academy of Management Review* 10 (4): 803–813.

Flavell, J. H. 1999. Cognitive development: Children's knowledge about the mind. *Annual Review of Psychology* 50 (1): 21–45.

Floyd, S. W., and B. Wooldridge. 1992. Middle management involvement in strategy and its association with strategic type: A research note. *Strategic Management Journal* 13: 153–167.

Forsyth, D. R. 2006. *Group Dynamics*. Thomson Wadsworth.

Frey, B. S., and R. Jegen. 2001. Motivation crowding theory. *Journal of Economic Surveys* 15 (5): 589–611.

Galbraith, J. R. 1973. *Designing Complex Organizations*. Addison-Wesley.

Galbraith, J. R. 2002. *Designing Organizations*. Jossey-Bass.

Galton, F. 1907. Vox populi. *Nature* 75: 450–451.

Guzzo, R. A., and M. W. Dickson. 1996. Teams in organizations: Recent research on performance and effectiveness. *Annual Review of Psychology* 47: 307–338.

Hackman, J. R., and G. R. Oldham. 1976. Motivation through the design of work: Test of a theory. *Organizational Behavior and Human Performance* 16: 250–279.

Harrison, D. D., and K. J. Klein, K. 2007. What's the difference? Diversity constructs as separation, variety, or disparity in organizations. *Academy of Management Review* 32 (4): 1199–1228.

Hinsz, V. B., R. S. Tindale, and D. A. Vollrath. 1997. The emerging conceptualization of groups as information processors. *Psychological Bulletin* 121 (1): 43–64.

Hollenbeck, J. R., D. S. DeRue, and R. Guzzo, R. 2004. Bridging the gap between I/O research and HR practice: Improving team composition, team training, and team task design. *Human Resource Management* 43 (4): 353–366.

Hollingshead, A. B. 2001. Cognitive interdependence and convergent expectations in transactive memory. *Journal of Personality and Social Psychology* 81 (6): 1080–1089.

Holmstrom, B. 1982. Moral hazard in teams. *Bell Journal of Economics* 13 (2): 324–340.

Hong, L., and S. E. Page. 2004. Groups of diverse problem solvers can outperform groups of high-ability problem solvers. *Proceedings of the National Academy of Sciences* 101 (46): 16385–16389.

Horwitz, S. K., and I. B. Horwitz. 2007. The effects of team diversity on team outcomes: A meta-analytic review of team demography. *Journal of Management* 33 (6): 987–1015.

Isenberg, D. J. 1986. Group polarization: A critical review and meta-analysis. *Journal of Personality and Social Psychology* 50 (6): 1141–1151.

Janis, I. L., and L. Mann. 1977. *Decision Making: A Psychological Analysis of Conflict, Choice, and Commitment*. Free Press.

Jenkins, G. D., A. Mitra, N. Gupta, and J. D. Shaw. 1998. Are financial incentives related to performance? A meta-analytic review of empirical research. *Journal of Applied Psychology* 83 (5): 777—787.

Joshi, A., and H. Roh. 2009. The role of context in work team diversity research: A meta-analytic review. *Academy of Management Journal* 52 (3): 599–627.

Kellermanns, F. W., J. Walter, C. Lechner, and S. W. Floyd. 2005. The lack of consensus about strategic consensus: Advancing theory and research. *Journal of Management* 31: 719–737.

Kerr, N. L., and S. E. Bruun. 1983. Dispensability of member effort and group motivation losses: Free-rider effects. *Journal of Personality and Social Psychology* 44 (1): 78–94.

Kerr, N. L., R. J. MacCoun, and G. P. Kramer. 1996. Bias in judgment: Comparing individuals and groups. *Psychological Review* 103 (4): 687.

Kidd, D. C., and E. Castano. 2013. Reading literary fiction improves theory of mind. *Science* 342 (6156): 377–380.

Knott, A. M. 2008. R&D/returns causality: Absorptive capacity or organizational IQ. *Management Science* 54 (12): 2054–2067.

Kogut, B., and U. Zander. 1992. Knowledge of the firm, combinative capabilities, and the replication of technology. *Organization Science* 3 (3): 383–397.

Kruglanski, A. W., I. Friedman, and G. Zeevi. 1971. The effects of extrinsic incentive on some qualitative aspects of task performance. *Journal of Personality* 39 (4): 606–617.

Krych-Appelbaum, M., J. B. Law, A. Barnacz, A. Johnson, and J. P. Keenan. 2007. "I think I know what you mean": The role of theory of mind in collaborative communication. *Interaction Studies* 8 (2): 267–280.

Lakhani, K. R., and R. G. Wolf. R. G. 2005. Why hackers do what they do. In *Understanding Motivation and Effort in Free/Open Source Software Projects*, ed. J. Feller et al. MIT Press.

Larson, J. R. 2009. *In Search of Synergy in Small Group Performance*. Psychology Press.

Latane, B., K. Williams, and S. Harkins. 1979. Many hands make light the work: The causes and consequences of social loafing. *Journal of Personality and Social Psychology* 37 (6): 822–832.

Laughlin, P. R. 1980. Social combination processes of cooperative problem-solving groups on verbal intellective tasks. *Progress in Social Psychology* 1: 127–155.

Laughlin, P. R., and A. L. Ellis. 1986. Demonstrability and social combination processes on mathematical intellective tasks. *Journal of Experimental Social Psychology* 22 (3): 177–189.

Lawrence, P. R., and J. W. Lorsch. 1967. *Organization and Environment : Managing Differentiation and Integration*. Harvard Business School Press.

Lazear, E. P. 2000. Performance pay and productivity. *American Economic Review* 90 (5): 1346–1361.

Lazear, E. P., and K. L. Shaw. 2007. Personnel economics: The economist's view of human resources. *Journal of Economic Perspectives* 21 (4): 91–114.

Lerner, J., and J. Tirole. 2005. The scope of open source licensing. *Journal of Law Economics and Organization* 21 (1): 20–56.

Lewis, K. 2004. Knowledge and performance in knowledge-worker teams: A longitudinal study of transactive memory systems. *Management Science* 50: 1519–1533.

Lewis, K., and B. Herndon. 2011. The relevance of transactive memory systems for complex, dynamic group tasks. Presented at Organization Science Winter Conference.

Liang, D. W., R. Moreland, and L. Argote. 1995. Group versus individual training and group performance: The mediating role of transactive memory. *Personality and Social Psychology Bulletin* 21 (4): 384–393.

Locke, E. A., C. C. Durham, J. M. L. Poon, and E. Weldon. 1997. "Goal setting, planning, and performance on work tasks for individuals and groups." In *The Developmental Psychology of Planning*, ed. S. L. Friedman and E. K. Scholnick. Erlbaum.

Locke, E. A., and G. P. Latham. 2006. New directions in goal-setting theory. *Current Directions in Psychological Science* 15 (5): 265–268.

Locke, E. A., K. N. Shaw, L. M. Saari, and G. P. Latham. 1981. Goal setting and task performance. *Psychological Bulletin* 90 (1): 125–152.

Lorenz, J., H. Rauhut, F. Schweitzer, and D. Helbing. 2011. How social influence can undermine the wisdom of crowd effect. *Proceedings of the National Academy of Sciences* 108 (22): 9020–9025.

Malone, T. W. 2004. *The Future of Work*. Harvard Business Review Press.

Malone, T. W., K. Crowston, J. Lee, B. Pentland, C. Dellarocas, G. Wyner, et al. 1999. Tools for inventing organizations: Toward a handbook of organizational processes. *Management Science* 45 (3): 425–443.

Malone, T. W., R. Laubacher, and C. Dellarocas. 2010. The collective intelligence genome. *MIT Sloan Management Review* 51 (3): 21–31.

Malone, T. W., J. Yates, and R. I. Benjamin. 1987. Electronic markets and electronic hierarchies. *Communications of the ACM* 30 (6): 484–497.

Manso, G. 2011. Motivating innovation. *Journal of Finance* 66 (5): 1823–1860.

March, J. G., and H. A. Simon. 1958. *Organizations*. Wiley.

Marks, M. A., J. E. Mathieu, and S. J. Zaccaro. 2001. A temporally based framework and taxonomy of team processes. *Academy of Management Review* 26 (3): 356–376.

Mayer, J. D., and P. Salovey. 1993. The intelligence of emotional intelligence. *Intelligence* 17 (4): 433–442.

McGrath, J. E. 1984. *Groups: Interaction and Performance*. Prentice-Hall.

Mesmer-Magnus, J. R., and L. A. DeChurch. 2009. Information sharing and team performance: A meta-analysis. *Journal of Applied Psychology* 94 (2): 535–546.

Milliken, F. J., and L. L. Martins. 1996. Searching for common threads: Understanding the multiple effects of diversity in organizational groups. *Academy of Management Review* 21 (2): 402–433.

Moreland, R. L., L. Argote, and R. Krishnan. 2002. Training people to work in groups. In *Theory and Research on Small Groups*, ed. R. S. Tindale et al. Springer.

Moreland, R. L., and L. Myaskovsky. 2000. Exploring the performance benefits of group training: Transactive memory or improved communication? *Organizational Behavior and Human Decision Processes* 82 (1): 117–133.

Nonaka, I. 1994. A dynamic theory of organizational knowledge creation. *Organization Science* 5 (1): 14–37.

Ocasio, W. 1997. Towards an attention-based view of the firm. *Strategic Management Journal* 18 (S1): 187–206.

Ocasio, W. 2011. Attention to attention. *Organization Science* 22 (5): 1286–1296.

Okhuysen, G. A., and B. A. Bechky. 2009. Coordination in organizations: An integrative perspective. *Academy of Management Annals* 3 (1): 463–502.

O'Leary-Kelly, A. M., J. J. Martocchio, and D. D. Frink. 1994. A review of the influence of group goals on group performance. *Academy of Management Journal* 37 (5): 1285–1301.

Peterson, C. C., V. P. Slaughter, and J. Paynter. 2007. Social maturity and theory of mind in typically developing children and those on the autism spectrum. *Journal of Child Psychology and Psychiatry* 48 (12): 1243–1250.

Powell, W. 1990. Neither market nor hierarchy: Network forms of organization. *Research in Organizational Behavior* 12: 295–336.

Premack, D., and G. Woodruff. 1978. Does the chimpanzee have a theory of mind? *Behavioral and Brain Sciences* 1 (04): 515–526.

Prendergast, C. 1999. The provision of incentives in firms. *Journal of Economic Literature* 37 (1): 7–63.

Reiter-Palmon, R., B. Wigert, and T. D. Vreede. 2012. Team creativity and innovation: The effect of group composition, social processes, and cognition. In *Handbook of Organizational Creativity*, ed. M. D. Mumford. Academic Press.

Roloff, K. S., A. W. Woolley, and A. Edmondson. 2011. The contribution of teams to organizational learning. In *Handbook of Organizational Learning and Knowledge Management*, second edition, ed. M. Easterby-Smith and M. Lyles. Wiley.

Sanders, G. S., and R. S. Baron. 1977. Is social comparison irrelevant for producing choice shifts? *Journal of Experimental Social Psychology* 13 (4): 303–314.

Saxe, R. 2009. Theory of mind (neural basis). In *Encyclopedia of Consciousness*, volume 2. Academic Press.

Senge, P. M., and J. D. Sterman. 1992. Systems thinking and organizational learning: Acting locally and thinking globally in the organization of the future. In *Transforming Organizations*, ed. T. Kochan and M. Useem. Oxford University Press.

Simon, H. A. 1947. *Administrative Behavior: A Study of Decision-Making Processes in Administrative Organizations*. Macmillan.

Stasser, G., and W. Titus. 1985. Pooling of unshared information in group decision making: Biased information sampling during discussion. *Journal of Personality and Social Psychology* 57: 67–78.

Steiner, I. 1972. *Group Process and Productivity*. Academic Press.

Sternberg, R. J., and W. Salter. 1982. Conceptions of intelligence. In *Handbook of Human Intelligence*, ed. R. J. Sternberg. Cambridge University Press.

Styles, E. 2006. *The psychology of attention*, second edition. Psychology Press.

Surowiecki, J. 2004. *The Wisdom of Crowds*. Random House.

Tesluk, P. E., J. L. Farr, and S. R. Klein. 1997. Influences of organizational culture and climate on individual creativity. *Journal of Creative Behavior* 31 (1): 27–41.

Thompson, J. D. 1967. *Organizations in Action*. McGraw-Hill.

Thorndike, E. L. 1898. Animal intelligence: An experimental study of the associative processes in animals. *Psychological Review* 2 (4): 1–109.

Toma, C., and F. Butera. 2009. Hidden profiles and concealed information: Strategic information sharing and use in group decision making. *Personality and Social Psychology Bulletin* 35 (6): 793–806.

Van de Ven, A. H., A. L. Delbecq, and R. Koenig. 1976. Determinants of coordination modes within organizations. *American Sociological Review* 41 (2): 322–338.

Van Knippenberg, D., and M. C. Schippers. 2007. Work group diversity. *Annual Review of Psychology* 58: 515–541.

Wageman, R. 1995. Interdependence and group effectiveness. *Administrative Science Quarterly* 40 (1): 145–180.

Wageman, R., and G. Baker. 1997. Incentives and cooperation: The joint effects of task and reward interdependence on group performance. *Journal of Organizational Behavior* 18 (2): 139–158.

Watson, A. C., C. L. Nixon, A. Wilson, and L. Capage. 1999. Social interaction skills and theory of mind in young children. *Developmental Psychology* 35 (2): 386.

Wegner, D. M. 1987. Transactive memory: A contemporary analysis of the group mind. In *Theories of Group Behavior*, ed. B. Mullen and G. R. Goethals. Springer.

Weingart, L. R. 1992. Impact of group goals, task component complexity, effort, and planning on group performance. *Journal of Applied Psychology* 77 (5): 682–693.

Weldon, E., and L. R. Weingart. 1993. Group goals and group performance. *British Journal of Social Psychology* 32: 307–334.

Williams, K. Y., and C. A. I. O'Reilly. 1998. Demography and diversity in organizations: A review of 40 years of research. *Research in Organizational Behavior* 20: 77–140.

Williamson, O. E. 1973. Markets and hierarchies. *American Economic Review* 63: 316–325.

Williamson, O. E. 1981. The economics of organization: The transaction cost approach. *American Journal of Sociology* 87 (3): 548–577.

Wilson, J. M., P. S. Goodman, and M. A. Cronin. 2007. Group learning. *Academy of Management Review* 32 (4): 1041–1059.

Wimmer, H., and J. Perner. 1983. Beliefs about beliefs: Representation and constraining function of wrong beliefs in young children's understanding of deception. *Cognition* 13 (1): 103–128.

Wittenbaum, G. M., S. I. Vaughan, and G. Stasser. 1998. Coordination in task-performing groups. In *Theory and Research on Small Groups*, ed. R. S. Tindale et al. Plenum.

Woolley, A. W. 2009a. Means versus ends: Implications of outcome and process focus for team adaptation and performance. *Organization Science* 20: 500–515.

Woolley, A. W. 2009b. Putting first things first: Outcome and process focus in knowledge work teams. *Journal of Organizational Behavior* 30: 427–452.

Woolley, A. W. 2011. Playing offense versus defense: The effects of team strategic orientation on team process in competitive environments. *Organization Science* 22: 1384–1398.

Woolley, A. W., J. B. Bear, J. W. Chang, and A. H. DeCostanza. 2013. The effects of team strategic orientation on team process and information search. *Organizational Behavior and Human Decision Processes* 122 (2): 114–126. doi:10.1016/j.obhdp.2013.06.002.

Woolley, A. W., C. F. Chabris, A. Pentland, N. Hashmi, and T. W. Malone. 2010. Evidence for a collective intelligence factor in the performance of human groups. *Science* 330 (6004): 686–688.

Woolley, A. W., M. E. Gerbasi, C. F. Chabris, S. M. Kosslyn, and J. R. Hackman. 2008. Bringing in the experts: How team composition and work strategy jointly shape analytic effectiveness. *Small Group Research* 39 (3): 352–371.

Woolley, A. W., J. R. Hackman, T. J. Jerde, C. F. Chabris, S. L. Bennett, and S. M. Kosslyn. 2007. Using brain-based measures to compose teams: How individual capabilities and team collaboration strategies jointly shape performance. *Social Neuroscience* 2: 96–105.

Law, Communications, Sociology, Political Science, and Anthropology

Editors' Introduction

Suppose that we asked all the people in the world to join together and spend their free time building a spaceship so that humanity could send a manned mission to Mars. It probably sounds preposterous—but a few years ago, so did the idea of a globally editable encyclopedia. Today, however, millions of people use Wikipedia, and it is has become the largest encyclopedia in history. Wikipedia's success challenged many traditional notions in law, economics, and sociology. How could its volunteer contributors have coordinated their activities? Why would they have dedicated so many hours to it without pay? Who is responsible for the result?

The following chapter on peer production synthesizes these questions to investigate activities associated with collective intelligence. It focuses especially on the activities of distributed volunteers: not just writing encyclopedias, but also remixing creative work and creating open-source software. Benkler, Shaw, and Hill walk through three main questions, one concerning governance, one concerning motivation, and one concerning quality. For each question, they encourage the reader to engage with a number of unanswered cross-disciplinary questions.

Capturing the ambition of the chapter that follows within a single social scientific discipline would be nearly impossible. The authors cover an impressive set of literatures and disciplines under the banner of peer production. They extract significant value from synthesizing across disciplines. Law seeks to understand how notions of policy and ownership affect our understanding of peer production. Sociology studies the social organization of these phenomena and brings large-scale analysis techniques to the table. Communication contributes both critical perspectives and empirical evidence.

These fields also have much to contribute to the study of collective intelligence beyond the scope of the chapter by Benkler and colleagues. Sociology, for example, helps us to understand the patterns, power, and structure of our social lives. One major area of research in sociology uses the tools of network science to understand how the connections between individuals can produce intelligent decisions and innovation. For example, Watts and Strogatz (1998) demonstrated how networks that are mainly connected in tight groups but with a small number of random links to remote nodes have surprising "small-world" properties. Many networks with such structures support quite intelligent outcomes, such as neural pathways and the American power grid.

Sociologists are also aware that clusters within a social network hold mainly redundant information (Granovetter 1973). However, individuals who can help bridge the gaps between these clusters (known as *structural holes*) play an important role in a network's ability to act on the knowledge distributed within its own members (Burt 1992). These large, macro-scale networks can even provide insight into micro-scale behaviors. For example, scientific collaboration networks have shed light on how teams of researchers, rather than solo authors, are increasingly shaping the research landscape (Wuchty, Jones, and Uzzi 2007).

A second relevant area of sociology seeks to understand what a group values and what it does not value. How much of the value that we place on songs, politicians, and institutions is socially constructed (e.g., we make it up), and how much is truly inherent in the objects? Work in which we analyze groups in terms of how well they evaluate the inherent worth of these items provides insights into the collective decision-making capabilities of the groups.

In one experiment that we include in the recommended readings below, the researchers set up an online music-listening site that was populated with songs by unknown musicians (Salganik, Dodd, and Watts 2006). Listeners were randomized into several self-contained universes in which they could see the play counts from other participants. While the most popular and the least popular songs remained relatively consistent across universes, the rest of the songs had no common ordering. Clearly, listeners were constructing much of the songs' value as the experiment ran by observing what songs other listeners thought were popular.

More broadly, sociologists aim to understand the degree of this social construction as opposed to objective constraints, as well as how these socially constructed values bubble and burst over time

(Zuckerman 2012). The sociology of social (e)valuation spreads not only to technical phenomena but also to broader social patterns, such as government performance and school progress (Lamont 2012).

The field of communications helps us understand how the media we use will influence and be influenced by collective activities. For example, Manuel Castells (1996) argues that access to information and networks is increasingly forming the basis of our society. In doing so, networked societies are experiencing new forces of political mobility and exclusion. These forces touch upon and impact collective intelligence activities. This has played out more recently in a thread of new media studies, many of them focusing on participatory culture. A striking example is how movie, game, and anime fans now work together to remix and extend their beloved fictional universes and give them new meaning (Jenkins 2006). These collective activities often produce results of strikingly high quality.

Within communications, the subfield of communication networks brings a macro-scale social scientific lens to these phenomena. Monge and Contractor's 2003 book *Theories of Communication Networks* lays a foundation for such work by connecting social theory to network structures and by demonstrating how to apply inferential statistics to networks. These methods have proved useful in studying many intelligent phenomena that occur over networks—for example, the conditions under which teams of scientists are most likely to make breakthroughs (Börner et al. 2010).

The field of political science helps us to understand how groups can govern themselves effectively. Political scientists debate the relationship between governing institutions and the most collectively intelligent or efficient outcomes. The field has built up a body of knowledge comparing alternative forms of democracy, such as parliamentary and presidential systems. One example of an outcome is that parliamentary systems must choose between being representative and being efficient (Shugart and Carey 1992).

An effective (or intelligent) government relies on active participants, and so political science has produced a mature string of work on voter participation. Social comparison and social desirability bias can be effective: in one experiment, some residents of neighborhoods in Michigan were sent postcards containing information about whether their neighbors had voted in the preceding election. The implication, of course, was that one's neighbors could know whether one had voted.

Turnout was much higher among people who had received these social-pressure mailings (Gerber, Green, and Larimer 2008).

A times, collective intelligence also relies on *collective action*, in which a group takes a united stance on a topic. However, in many cases there is a cost to individuals of taking action, and it is easier to instead "free ride" on the efforts of others if those efforts will result in a public good available even to those who did not participate. This results in a *collective-action problem*: everyone may privately want to effect a change, but no one wants to pay the costs of starting the change unless they know that many others will join them and thus dilute their individual cost (Olson 1965). Threshold models of collective action describe how these efforts tend to have a level of publicly expressed support that encourages everyone to join (Granovetter 1978; Kuran 1991).

The following chapter gives just a taste of research in a wide range of social science fields that offers insights into ways of analyzing and designing collectively intelligent systems.

References

Börner, Katy, Noshir Contractor, Holly Falk-Krzesinski, et al. 2010. A multi-level systems perspective for the science of team science. *Science Translational Medicine* 2 (49): 49cm24.

Burt, Ronald S. 1992. *Structural Holes: The Social Structure of Competition.* Harvard University Press.

Castells, Manuel. 1996. *The Rise of the Network Society: The Information Age: Economy, Society, and Culture.* Wiley.

Gerber, Alan S., Donald P. Green, and Christopher W. Larimer. 2008. Social pressure and voter turnout: Evidence from a large-scale field experiment. *American Political Science Review* 102 (1): 33–48.

Granovetter, Mark S. 1973. The strength of weak ties. *American Journal of Sociology* 78 (6): 1360–1380.

Granovetter, Mark. 1978. Threshold models of collective behavior. *American Journal of Sociology* 83 (6): 1420–1443.

Jenkins, Henry. 2006. *Convergence Culture.* NYU Press.

Kuran, Timur. 1991. Now out of never: The element of surprise in the East European Revolution of 1989. *World Politics* 44 (1): 7–48.

Lamont, Michèle. 2012. Toward a comparative sociology of valuation and evaluation. *Annual Review of Sociology* 38 (1): 201–221.

Monge, Peter, and Noshir Contractor. 2003. *Theories of Communication Networks.* Oxford University Press.

Olson, Mancur. 1971. *1965. The Logic of Collective Action: Public Goods and the Theory of Groups.* Revised edition, Harvard University Press.

Salganik, Matthew J., Peter Sheridan Dodds, and Duncan J. Watts. 2006. Experimental study of inequality and unpredictability in an artificial cultural market. *Science* 311 (5762): 854–856.

Shugart, Matthew Soberg, and John M. Carey. 1992. *Presidents and Assemblies: Constitutional Design and Electoral Dynamics.* Cambridge University Press.

Watts, Duncan, and Steven H. Strogatz. 1998. Collective dynamics of "small-world" networks. *Nature* 393 (6684): 440–442.

Wuchty, Stefan, Benjamin F. Jones, and Brian Uzzi. 2007. The increasing dominance of teams in production of knowledge. *Science* 316 (5827): 1036–1039.

Zuckerman, Ezra W. 2012. Construction, concentration, and (dis)continuities in social valuations. *Annual Review of Sociology* 38:223–245.

Recommended Readings

Yochai Benkler. 2006. *The Wealth of Networks: How Social Production Transforms Markets and Freedom.* Yale University Press.

This foundational book developed Yochai Benkler's explorations of peer production, which he termed the networked information economy. In it he drew together a wide range of disciplines and legitimized the study of processes of peer production. In doing so, he laid a theoretical basis for many fields' subsequent explorations.

Aniket Kittur and Robert E. Kraut. 2008. Harnessing the wisdom of crowds in Wikipedia: Quality through coordination. In *Proceedings of the 2008 ACM Conference on Computer Supported Cooperative Work.* ACM.

Exploiting natural variation in the number of editors of Wikipedia pages and the coordination strategies they use, Kittur and Kraut seek to understand how distributed volunteers can act most effectively. They find that adding more contributors to an article helped only when the editors coordinated effectively, and that larger groups of editors actively diminished an article's quality if the editors didn't coordinate effectively. They suggest that implicit coordination, in which authors took action rather than spend all their time discussing, was central to the eventual success of these articles.

Michael Kearns, Siddharth Suri, and Nick Montfort. 2006. An experimental study of the coloring problem on human subject networks. *Science* 313 (5788): 824–827.

Many kinds of collective intelligence involve individual actors who are connected by various network structures, and the field of network science has extensively studied the structural properties of such networks. But since much of that work involves naturally occurring networks, it is difficult to determine how much the behavior of these actors is influenced by the structure of the network itself and how much by other factors. This paper shows how experiments with artificially constructed networks can begin to untangle these factors. For instance, the authors found that providing more information to group members can make it easier or harder for them to solve group decision problems, depending on the structure of the network.

Galen Pickard, Wei Pan, Iyad Rahwan, Manuel Cebrian, Riley Crane, Anmol Madan, and Alex Pentland. 2011. Time-critical social mobilization. *Science* 334 (6055): 509–512.

The Defense Advanced Research Projects Agency (DARPA) ran a highly publicized challenge in which it launched ten red balloons across the United States. The first team to submit the correct locations of all ten balloons won a significant cash prize. A team at MIT developed a recursive incentive strategy such that each contributor would receive a percentage of the money if someone recruited by one of them found a balloon. That team grew virally and won the challenge in under nine hours. Pickard et al. report on the team's experience and on the recursive incentive mechanism it used.

Matthew J. Salganik, Peter Sheridan Dodds, and Duncan J. Watts. 2006. Experimental study of inequality and unpredictability in an artificial cultural market. *Science* 311 (5762): 854–856.

This already classic paper demonstrates empirically how much unpredictability in group decision making can result from social influences within a group. The researchers created an open website at which people listened to music. The listeners were split randomly into one of several "universes," each with its own distinct download counts. Each "universe" thus developed its own set of rankings of the popularity of songs. The resulting ratings of the same songs were extremely different across "universes," suggesting that the group members' ratings were very strongly influenced by the vagaries of how early raters in each world rated the songs.

Peer Production: A Form of Collective Intelligence

Yochai Benkler, Aaron Shaw, and Benjamin Mako Hill

Wikipedia has mobilized a collective of millions to produce an enormous, high-quality encyclopedia without traditional hierarchical organization or financial incentives. More than any other twenty-first-century collaborative endeavor, Wikipedia has attracted the attention of scholars in the social sciences and law both as an example of what collective intelligence makes possible and as an empirical puzzle. Conventional thinking suggests that effective and successful organizations succeed through hierarchical control and management, but Wikipedia seems to organize collectively without either. Legal and economic common sense dictates that compensation and contracts are necessary for individuals to share the valuable products of their work, yet Wikipedia elicits millions of contributions without payment or ownership. Intuition suggests that hobbyists, volunteers, and rag-tag groups will not be able to create information goods of sufficient quality to undermine professional production, but contributors to Wikipedia have done exactly that. Wikipedia should not work, but it does.

The gaps between conventional wisdom about the organization of knowledge production and the empirical reality of collective intelligence produced in "peer production" projects such as Wikipedia have motivated research on fundamental social scientific questions about the organization and the motivation of collective action, cooperation, and distributed knowledge production. Historically, researchers in such diverse fields as communication, sociology, law, and economics have argued that effective human systems organize people through a combination of hierarchical structures (e.g., bureaucracies), completely distributed coordination mechanisms (e.g., markets), and social institutions of various kinds (e.g., cultural norms). However, the rise of networked systems and online platforms for collective intelligence has upended many of the assumptions and the findings of this earlier research. In the process, online collective intelligence systems have generated data sources for the study of social organization and behavior at unprecedented scale and granularity. As a result, social scientific research on collective intelligence continues to grow rapidly.

Wikipedia demonstrates how participants in many collective intelligence systems contribute valuable resources without the hierarchical bureaucracies or strong leadership structures common to state agencies or firms and in the absence of clear financial incentives or rewards. Consistent with this example, foundational social scientific research relevant to understanding collective intelligence has focused on three central concerns: explaining the organization and governance of decentralized projects, understanding the motivation of contributors in the absence of financial incentives or coercive obligations, and evaluating the quality of the products generated through collective intelligence systems.

Much of the research on collective intelligence that has been done in the social sciences has focused on peer production. Following Benkler (2013), we define peer production as a form of open creation and sharing performed by online groups that set and execute goals in a decentralized manner, harness a diverse range of participants' motivations (particularly non-monetary motivations), and separate governance and management relations from exclusive forms of property and relational contracts (i.e., projects are governed as open commons or common property regimes, and organizational governance utilizes combinations of participatory, meritocratic, and charismatic models rather than proprietary or contractual models). Peer production is the most significant organizational innovation to have emerged from Internet-mediated social practices, is among the most visible and important examples of collective intelligence, and is a central theoretical frame used by social scientists and legal scholars of collective intelligence.

Peer production includes many of the largest and most important collaborative communities on the Internet. The best-known instances of peer production include collaborative free/libre and open-source software (FLOSS), such as the GNU/Linux operating system, and free culture projects, such as Wikipedia. GNU/Linux is widely used in servers, mobile phones, and embedded systems such as televisions. Other FLOSS systems, such as the Apache Web server, are primary Internet utilities. Through participation in FLOSS development, many of the largest global technology companies have adopted peer production as a business strategy. Explicitly inspired by FLOSS, Wikipedia has been the subject of several books and of more than 6,200 academic articles.1 Wikipedia is among the top six websites in the world, with more than half a billion unique viewers per month.2 The success of FLOSS and Wikipedia has coincided with a variety of peer-production projects in an array of industries, including those that arose in stock photography, videos, and travel guides. For-profit and non-profit organizations have found ways to organize their products through peer production to overcome competition from more traditional market-based and firm-based approaches.

Although peer production is central to social scientific and legal research on collective intelligence, not all examples of collective intelligence created in online systems entail peer production. First, collective intelligence can involve centralized control over the setting of goals and the execution of tasks. For example, InnoCentive provides a platform for companies to distribute searches for solutions to difficult problems; however, the problems are defined within the boundaries of InnoCentive's client firms ("seekers"), and solutions are returned to seekers as complete products created by individuals or by teams. Peer production, in contrast, decentralizes both the setting of goals and the execution of tasks to networks of individuals or communities (Brabham 2013). Second, many collective intelligence platforms, such as the ESP Game and Amazon's Mechanical Turk labor market, focus on optimizing systems around a relatively narrow set of motivations and incentives—ludic and financial in ESP and Mechanical Turk respectively. Peer-production projects involve a broad range of incentives. Third, collective intelligence often occurs within firms in which participants are bound by the obligation of contracts and in contexts where the resources used or the products of collective effort are managed through exclusive property rights. Peer production exists without these structures.

Since the late 1990s, when FLOSS first acquired widespread visibility and broad economic relevance, peer production has attracted scholarly attention. The earliest work sought to explain the surprising success of peer production and to draw distinctions between peer production and traditional models of firm-based production. Although many of the first analyses focused on software, the concept of peer production emerged as the most influential attempt to situate FLOSS as an instance of a broader phenomenon of online cooperation (Benkler 2001, 2002). A number of scholars (e.g., von Hippel and von Krogh 2003; Weber 2004) then adopted the concept of peer production to emphasize the comparative advantages of Internet-based cooperation as a mode of innovation and information production.

In the rest of this chapter, we describe the development of the academic literature on peer production and collective intelligence in three areas: organization, motivation, and quality. In each area, we introduce foundational work—primarily earlier scholarship that sought to describe peer production and establish its legitimacy. Subsequently, we characterize work, usually more recent, that seeks to pursue new directions and to derive more nuanced analytical insights. Because FLOSS and Wikipedia have generated the majority of the research on peer production to date, we focus on research analyzing those efforts. For each theoretical area, we briefly synthesize foundational work and new directions and describe some of the challenges for future scholarship. Our themes are not intended to encompass the complete literature on peer production, nor is our periodization. Instead, they echo the areas of research that speak most directly to the literature on collective intelligence and locate peer production within that broader phenomenon. We conclude with a discussion of several issues that traverse our themes and implications of peer-production scholarship for research on collective intelligence more broadly.

Organization

Early scholars of peer production were struck by the apparent absence of formal hierarchies and leadership structures. Indeed, peer-production communities perform such "classical" organizational functions as coordination, division of labor, recruitment, training, norm creation and enforcement, conflict resolution, and boundary maintenance, but do so in the absence of many of the institutions associated with more traditional organizations. In Williamson's (1985) terms, peer-production communities thrive despite a relative absence of bureaucratic structure, exclusive property rights, and relational contracts. However, as peer-production communities have aged, some have acquired increasingly formal organizational attributes, including bureaucratic rules and routines for interaction and control. The challenge for early peer-production research was to elaborate an account of the characteristics that differentiate peer production from traditional organizations. More recent work has sought to understand when the organizational mechanisms associated with peer production are more and less effective and how organizational features of communities change over time.

Foundational Work

A diverse and interdisciplinary group of scholars provided initial explanations for how and why peer-production communities function. Early organizational analyses of peer production focused on discussions of how peer production compared with modes of organization in firms, states, and markets. Descriptive work also characterized patterns and processes of governance and leadership within some of the most prominent peer-production communities.

Initial scholarship on the organization of peer production emphasized the presence of very low transactions costs associated with contributing and diverse participant motivations to define peer production as distinct from, bureaucracy, firms, and markets. Specifically, Benkler (2002, 2004, 2006) focused on the role of non-exclusive property regimes and more permeable organizational boundaries (i.e., organization without relational contracts) to explain how peer production arrived at more efficient equilibria for the production of particular classes of informational goods. Peer production could, Benkler argued, outperform traditional organizational forms under conditions of widespread access to networked communications technologies, a multitude of motivations driving contributions, and non-rival information capable of being broken down into granular, modular, and easy-to-integrate pieces. Other foundational accounts, including those of Moglen (1999) and Weber (2004), attended to the emergence of informal hierarchies and governance arrangements within communities. This early work relied on case studies and on descriptive characterizations of how these arrangements functioned in comparison with traditional bureaucratic practices within organizations historically tasked with the production of information goods. At the extreme, peer-production communities were treated as a new category of "leaderless" social systems relying on informal norms such as reciprocity and fairness to solve complex problems *without* organizations.

Much of the earliest empirical research (see, e.g., von Krogh et al. 2003; Kelty 2008) inductively sought to describe the structure and organization of FLOSS communities by, for example, characterizing contributions across FLOSS projects through reference to a core and a periphery of participants (see, e.g., Lakhani 2006). Other work (see, e.g., O'Mahony and Bechky 2008; West and O'Mahony 2005; Spaeth et al. 2010) focused on describing the interactions between formal organizations and FLOSS communities in a variety of contexts. Crowston et al. (2010) provide a detailed overview of the scholarly literature on FLOSS organization and practice, focusing on work that we characterize as foundational.

A large body of descriptive work has also sought to characterize the organization of Wikipedia and the practices of its contributors. A series of qualitative accounts (e.g., Bryant et al. 2005; Forte and Bruckman 2008; Pentzold 2011) have examined the norms of participation and consensus among Wikipedians. Reagle (2010) and Jemielniak (2014) have provided the most influential book-length scholarly descriptions of Wikipedia and its processes. Another influential approach has employed data mining to build inductive, quantitative descriptions of Wikipedia's organization (Ortega 2009; Priedhorsky et al. 2007; Viégas et al. 2007; Yasseri et al. 2012). A growing body of related research has characterized Wikipedia's governance systems (Butler et al. 2008; Konieczny 2009), the emergence of leadership within the Wikipedia community (Forte et al. 2009; Zhu et al. 2011, 2012), mechanisms through which new community members are incorporated and socialized (Antin et al. 2012; Antin and Cheshire 2010; Halfaker et al. 2011), and the ways in which bureaucratic decisions (for example, about promotions) are made (Burke and Kraut 2008; Leskovec et al. 2010).

Although they are a relative minority, foundational scholarly descriptions of organizational elements of peer-production communities other than FLOSS and Wikipedia have emphasized similar concerns. For example, analyses of Slashdot's system of distributed moderation provided an early illustration of how recommendation and filtering constituted an informal mode of governance and supported a division of labor on the site (Lampe and Resnick 2004; Lampe et al. 2007). This research built on earlier work (e.g., Fisher et al. 2006) showing that contributors to bulletin boards, newsgroups, forums, and related systems adopted durable "social roles" through their patterns of contribution—an approach that was subsequently applied to Wikipedia by Welser et al. (2011) and McDonald et al. (2011).

New Directions
More recent work on organizational aspects of peer production has begun to question the "stylized facts" that prevailed in earlier research. Some of this work compares projects and routines to understand the relationship between organizational attributes of communities and their outputs. Other work engages in interventions and field experiments to test and improve organizational operations in peer-production communities. Many of the best examples of this newer work have

incorporated collections of peer-production communities in compara-tive analyses. Some of these studies also articulate connections between peer-production research and theories from other organizational fields, including academic studies of firms, states, political parties, social movement organizations, and civic associations.

Although some of the earliest theories of the organization of peer production celebrated it as non-hierarchical, more recent work (e.g., Kreiss et al. 2011) has questioned both the putative lack of hierarchy and its purported benefits. Several empirical studies (see, e.g., Hindman 2008; Wu et al. 2009) have emphasized how feedback loops of attention and cumulative advantage can perpetuate starkly unequal distribu-tions of influence and hierarchy. Healy and Schussman (2003) have gone as far as to propose that hierarchy may even contribute to peer production's success, and Keegan and Gergle (2010) and Shaw (2012) have documented gate-keeping behavior in peer production and sug-gested that it may benefit projects. Several studies (e.g., Loubser 2010; Shaw and Hill 2014) have looked at wikis and shown that governance and hierarchies tend to become more pronounced as peer-production projects mature. O'Mahony and Ferraro (2007) argued that FLOSS proj-ects can depend on and reflect existing firm-based hierarchies. This body of scholarship does not refute the idea that peer production is less hierarchical than alternative forms of organization or the idea that it uses novel governance structures; however, the authors suggest a dif-ferent model of peer production than the stylized anti-hierarchies depicted in earlier work.

Although there is enormous variation in organizational form among peer-production projects, until recently there had been very few sys-tematic comparisons across communities. An important step toward such cross-organizational comparisons has been the creation of large data sets drawn from many communities. The FLOSSmole project has facilitated recent comparative scholarship by assembling and publish-ing a series of cross-project data sets in the area of software, including a widely used data set from the FLOSS hosting website SourceForge (Howison et al. 2006). In the most in-depth comparative study of FLOSS, Schweik and English (2012), building on work by Ostrom (1990), compared characteristics of FLOSS projects' organization, resources, governance, and context, using the FLOSSmole data set of SourceForge projects. These comparative studies have revealed not only that large collaborative projects are extraordinarily rare but also that many projects generating widely downloaded FLOSS resources are, in fact, not peer production at all but rather the products of firms or individuals.

Similar cross-organization studies in other areas of peer production, or studies comparing across different types of peer production, have remained challenging and rare. One difficulty with comparative work across organizations, in general, is designing research capable of supporting inference into the causes of organizational success and failure. Historically, failed attempts to build organizations do not enter into research data sets or attract scholarly attention. Two studies of peer production that have captured these failures are that of Kittur and Kraut (2010), who drew from a data set of 7,000 peer-production wikis from the hosting firm Wikia, and that of Shaw and Hill (2014), who used an updated version of that data set with ten times as many wikis. Others, including Ransbotham and Kane (2011), Wang et al. (2012), and Zhu et al. (2011), have tried to look across wikis by considering variation in sub-organizations within Wikipedia. Fuster Morell (2010) described a comparative case study of several hundred peer-production projects active in Catalonia. Despite these exceptions, there exist very few publicly available large-scale comparative data sets for types of peer-production projects outside of FLOSS.

Another path forward in organizational research in peer production lies in the use of field experiments and organizational interventions. Using a series of examples from peer production, Kraut and Resnick (2012) drew heavily on experimental results to form principles of community design. Halfaker et al. (2013b) reported experiments by the Wikimedia Foundation to alter the organizational structure around participation. Luther et al. (2013) analyzed the effects of a new system intended to support peer-produced animations through more effective allocation of leaders' time and effort. By structuring design changes as experiments, these studies make credible causal claims about the relationship of organizational structure and project outcomes that previous work had struggled to establish. By intervening in real communities, these efforts achieve a level of external validity that lab-based experiments cannot.

Organizational analysis of peer production remains critical future work, as an increasing number of organizations perform their activities in computer-mediated environments using the tools, techniques, and resources of peer production. As a result, a science of the mechanisms, procedures, and techniques necessary for the effective production of networked informational resources is increasingly important in various spheres of activity. By challenging stylized conceptions of peer-production organization, through comparative work, and through causal inference, future peer-production research can continue to deepen and extend foundational insights. Future work can also begin to address questions of whether, how, and why peer-production systems have transformed some existing organizational fields more profoundly than others. An empirically informed understanding of when and where organizational practices drawn from peer production provide efficient and equitable means to produce, disseminate, and access information can provide social impact beyond the insights available through the study of any individual community.

Motivation

A second quality of peer production that challenged conventional economic theories of motivation and cooperation was the absence of clear extrinsic incentives such as monetary rewards. Traditional economic explanations of behavior rely on the assumption of a fundamentally self-interested actor mobilized through financial or other incentives. In seeking to explain how peer-production projects attract highly skilled contributors without money, much of the literature on peer production has focused on questions of participants' motivation. In the earliest work on motivation in the context of FLOSS, an important goal was simply explaining how a system abandoning financial incentives could mobilize participants to freely share their code and effort. Even as it became apparent that peer production attracted contributors motivated by various non-monetary factors, work in economics focused on whether it was feasible to collapse these motivations into relatively well-defined models of self-interest. A newer wave of work has stepped back from that approach and has sought to explain how multiple motivational "vectors" figure in the creation of common-pool resources online—an approach that underscores an advantage of peer production in its capacity to enable action without requiring translation into a system of formalized, extrinsic carrots and sticks. Systems better able to engage self-motivated action will be better able to attract decentralized discovery of projects, resources, and solutions that are not mediated by prices or by command-and-control hierarchies of obligation (Osterloh et al. 2002).

Foundational Work

In the earliest research on motivation in peer production, Ghosh (1998) and Lerner and Tirole (2002) used case studies of FLOSS projects to assert that developers' motivations could be easily assimilated into standard economic models. Frequently cited as motivations in foundational work by these authors, by von Hippel and von Krogh (2003), by von Krogh (2003), and by others were the use value of the software to the contributing developer; the hedonic pleasure of building software; the increased human capital, reputation, or employment prospects; and social status within a community of peers. Other early accounts analyzing examples of peer production beyond FLOSS suggested additional motivations. For example, Kollock (1999) emphasized reciprocity, reputation, a sense of efficacy, and collective identity as salient social psychological drivers of contribution to online communities and forums. Early studies of Wikipedia contributors suggested that Wikipedians edited in response to a variety of motives, many of which were social psychological in character (Beenen et al. 2004; Forte and Bruckman 2008; Nov 2007; Rafaeli and Ariel 2008; Panciera et al. 2009). In studies of peer-to-peer file sharing networks and online information sharing communities, Cheshire (2007) and Cheshire and Antin (2008) emphasized social psychological motivation and demonstrated that peer feedback could provide a mechanism for activating reputational concerns and encouraging additional contributions. Benkler (2002, 2006) argued that the combination of many incentives created a potentially fragile interdependence of motivations in peer production. In anthropological accounts of FLOSS, Coleman (2012) and Coleman and Hill (2004) suggested that political ideologies of freedom play an important role in FLOSS production. Von Krogh et al. (2012) published a comprehensive survey of work characterizing motivations in open-source software.

Much of the scholarship on individual motivation in FLOSS, Wikipedia, and other communities has relied on surveys. Some of the most influential survey research on FLOSS was conducted by Rishab Ghosh and colleagues (e.g., Ghosh and Prakash 2000; Ghosh et al. 2002; Ghosh 2005). Other such research included the Boston Consulting Group's survey of hackers (Lakhani and Wolf 2005), and the FLOSS-US study (David and Shapiro 2008). Lakhani and Wolf (2005) emphasized the self-reported motivations of intellectual stimulation, hedonic gain, and skills building, whereas Ghosh et al. (2002) found reciprocity and skills development as the main motivation—a finding that Hars and Ou (2002) found was supported even among contributors who earned a living from their work on FLOSS. Ghosh and colleagues at the United Nations University in Maastricht also collaborated with the Wikimedia Foundation to conduct an early survey of Wikipedia editors and readers (Glott et al. 2010); that survey, like several subsequent surveys of Wikipedia editors, found that Wikipedia contributors likewise identified many different reasons for participation. Research in remixing communities has suggested a similarly diverse range of motivation. Despite their differences in emphasis, scope, and genre, all these surveys support the claim that motivations in peer production are diverse and heterogeneous.

A related body of early observational research into the motivations of participants in peer production described individuals' self-selection into particular social roles within projects and communities (Fisher et al. 2006). For example, Shah (2006) found that developers motivated by use value contributed more to what she called "gated" source communities, whereas developers who were motivated primarily by fun or pleasure contributed primarily to more purely peer-production-based projects. Using contributions to differently licensed FLOSS projects, Belenzon and Schankerman (2008) found that contributors were heterogeneous in their motivational profiles, but also argued that contributors self-sorted among projects. For example, contributors who were more responsive to extrinsic motivations like reputation and employment tended to contribute more to corporate-sponsored projects. Similarly, studies of Slashdot, Wikipedia, and other free culture communities suggest that participants' behavior patterns reflect the sorts of editorial and coordination tasks that are needed to produce relevant news and encyclopedic content (Lampe et al. 2007, 2010; Welser et al. 2011; Viégas et al. 2007; Kriplean et al. 2008; McDonald et al. 2011).

In other foundational research on motivation in peer production, Lerner and Schankerman (2010) and others have explored why organizations, firms, and governments, rather than individual users, choose to participate in open-source software development. In this vein, Schweik and English (2012) discuss firm-level motivations, including the rate of innovation, the capacity to collaborate with other firms, and avoiding dependence on sole-source providers. Another major motivation for organizations is that participating in peer production could enable firms to develop in-house expertise in a tacit-knowledge-rich innovation system and to increase their absorptive capacity (King and Lakhani 2011).

New Directions

Recent empirical studies of motivation in peer-production communities have drawn increasingly on observational data and field experiments. Much of this work has supported the finding that a mixture of motivations attract participants to contribute in peer production. However, a growing number of studies also suggest that these motives interact with one another in unpredictable ways and, as a result, are vulnerable to "crowding out" when the introduction of extrinsic incentives undermines intrinsic motivation (Frey and Jegen 2001). Robust evidence of varied participation patterns and motivations of contributors to peer-production communities has given rise to other explanations of why peer producers do what they do. In particular, these newer accounts have focused on social status, peer effects, prosocial altruism, group identification, and related social psychological dimensions of group behavior.

Recent work on Wikipedia and related peer-production projects has made use of observational data as well as data from field and laboratory experiments. Two important insights provided by some of this newer work are that contributors act for different reasons and that theories based on a single uniform motivational model are likely to mischaracterize the motivational dynamics. For example, Restivo and van de Rijt (2012) randomly distributed "barnstars" (informal awards that any Wikipedian can give to another in recognition of contributions) to a sample of editors with high numbers of edits and found that the awards caused an increase in subsequent contributions. In a follow-up study, Restivo and van de Rijt (2014) observed increases in subsequent contributions only among the most active Wikipedians who had received barnstars. Similarly, Hill et al. (2012) used observational data to compare award recipients who displayed their barnstars as a public signal of accomplishment with those who did not and found that the "signalers" edited more than the "non-signalers." Algan et al. (2013) combined observational data on the contribution history of 850 Wikipedians with the performance of the same individuals in a battery of laboratory social dilemmas. They found that that a preference for reciprocity (measured by contributions to public goods games) and for trustworthy behavior (measured in a trust game) predict an increase in contributions up to the median editor, but that a preference for reciprocity no longer predicted contributions for editors above the median level. Instead, a preference for social signaling (identified partly using the data set of Hill et al.) predicted above-median contributions, but the high contributors who were signalers did not exhibit particularly prosocial behavior in a laboratory setting. Algan et al. also found that a lab-measured taste for altruism was not associated with increased contributions to Wikipedia. In field experiments conducted by affiliates of the Wikimedia Foundation, Halfaker et al. (2013b) demonstrated that a feedback tool could elicit increased contributions among new editors so long as the cost (in terms of effort) of contribution remained low. Antin et al. (2012) combined longitudinal and survey data from a cohort of new Wikipedia contributors to show that differential patterns of contribution—as well as different reasons for contributing—emerged early in individual editors' contribution histories.

Evidence from this newer body of research shows that motivations differ *within* contributors and that different contributors have different mixes of motivations. This depth and variety makes the problem of designing incentives for peer-production systems more complex than foundational work suggested. Such complexity is further compounded by the fact that the effect of any design intervention (e.g., the introduction of a system of reward or punishment) will necessarily focus on harnessing a particular motivational driver and will necessarily be non-separable from its effects on other motivational drivers. Experimental and observational data have exhaustively documented that the effects of the standard economic incentives (e.g., payment or punishment) are not separable from their effects on social motivational vectors (Bowles and Hwang 2008; Bowles and Polania-Reyes 2012; Frey and Jegen 2001; Frey 1997). Similarly, social psychological motivations do not offer a panacea: although evidence published by Willer (2009), by Restivo and van de Rijt (2012), and by others indicates that the conferral of social status can provide an effective mechanism for eliciting prosocial or altruistic behavior under many conditions, more recent work by Restivo and van de Rijt (2014) suggests that the very same status-based incentives can generate diminishing returns beyond a certain point and may even hurt in some situations.

Resolving the tensions between different motivations and incentives presents a design challenge for peer-production systems and other collective platforms. The complex interdependence of motivations, incentive systems, and the social behaviors that distinct system designs elicit has led Kraut and Resnick (2012) to call for evidence-based social design and Benkler (2009, 2011) to call for cooperative human system design. Research on FLOSS by Alexy and Leitner (2011) has demonstrated that it is sometimes possible, with the appropriate normative framing, to combine paid and unpaid contributions without crowding out intrinsic motivation. However, we are not aware of any comparable studies of successful integration of material and prosocial rewards in other areas of peer production. Indeed, Weblogs Inc. and other websites have tried and failed to augment pure peer-production systems by offering material rewards to top contributors. With habituation and practice, internalized prosocial behaviors may lead people to adopt a more or less cooperative stance in specific contexts on the basis of their interpretation of the appropriate social practice and its coherence with their self-understanding of how to live well (von Krogh et al. 2012; Benkler and Nissenbaum 2006). Similarly, practice and experience may shift individuals' proclivity to cooperate and make the work of designers or organizational leaders easier. Antin and Cheshire (2010) have argued that motivations can shift over time because certain forms of activity not visible in contribution logs (for example, reading Wikipedia) can both incentivize others to contribute and constitute an essential step toward participation.

Observational work that estimates the effects of motivational drivers across real communities offers an especially promising avenue for future work. In this vein, planned and unplanned design changes to communities can provide "natural" experiments and may present particularly useful opportunities for researchers and community managers to evaluate changes to existing incentive systems. For example, Zhang and Zhu (2011) treated the Chinese government's decision to block Wikipedia as a natural experiment and estimate the effect of changes in "group size" on motivation to contribute among non-blocked users. Hill (2011) used a change to the remixing site Scratch to estimate the causal effect of a newly introduced reputational incentive. Because most peer-production systems are facilitated by software-defined Web applications, changing the fundamental rules of engagement within peer-production projects is simple and common. The diffusion of such techniques as A/B testing and quasi-experimental data analysis thus provides an enormous untapped pool of opportunities to study motivation.

Peer production successfully elicits contributions from diverse individuals with diverse motivations—a quality that continues to distinguish it from similar forms of collective intelligence. Although an early body of literature helped establish and document a complex web of motivations, more recent studies have begun to disentangle the distinct motivational profiles among contributors across cultures and communities, the interactions between motivational drivers and motivational profiles, and the design choices that can elicit and shape contribution.

Quality

A third kind of research into peer production focuses on the quality of peer production's outputs. The capacity of peer-production communities such as FLOSS and Wikipedia to create complex, technical products whose quality and scale rival and often outcompete the products of professional workforces in resource-rich firms has drawn both popular and academic interest. Indeed, the presence of novel organizational structures and unusual approaches to incentivizing contributors may have been uninteresting to many scholars of peer production if the products had been consistently of low quality. However, by the late 1990s GNU/Linux, Apache, and some other FLOSS projects were widely considered to be of higher quality than proprietary alternatives (i.e., proprietary UNIX operating systems and proprietary Web servers such as Microsoft's IIS, respectively). Other peer-produced goods have also equaled or overtaken their direct proprietary competitors. Recent work has begun to explore more deeply the different dimensions along which quality can be conceptualized and measured. This new scholarship has given rise to a more nuanced understanding of the different mechanisms through which high-quality resources arise, and founder, in peer production.

Foundational Work

FLOSS has been described—e.g., by Weber (2004)—as built upon a process that will inherently lead to high-quality outputs. A prominent FLOSS practitioner and advocate, Eric Raymond (1999), used the example of the Linux kernel to coin what he called Linus' Law—"with enough eyeballs, all bugs are shallow"—suggesting that FLOSS projects, by incorporating contributions from a large number of participants, would have fewer bugs than software developed through closed models. The mission statement of Open Source Initiative, the non-profit advocacy organization founded by Raymond in 1998, argues that "the promise of open source is better quality, higher reliability, more flexibility, lower cost, and an end to predatory vendor lock-in"3 and is the cited inspiration for many early academic studies of FLOSS (e.g., von Krogh and von Hippel 2006). Practitioner-advocates, FLOSS entrepreneurs, and scholars have each assumed that the peer-produced nature of FLOSS carried inherent quality benefits.

Raymond's argument for the superiority of FLOSS's development methodology gained currency as FLOSS emerged from relative obscurity in the late 1990s and began to be adopted by individuals and firms. The GNU/Linux operating system became peer production's most widely discussed success, and remains in wide use today. Recognizing the value produced through peer production, many established companies, including Google, HP, and Oracle, followed IBM's early lead and adopted FLOSS in major parts of their technology business. For example, Google made a strategic decision to develop Android as FLOSS, which allowed the product to catch up to and then overtake Apple's iOS as the dominant operating system for smart phones. Red Hat and other FLOSS-focused start-ups have been among the fastest-growing technology companies during the first ten years since its founding (Savitz 2012). FLOSS has also been particularly important in the creation of Internet infrastructure. In the cases of Web servers and server-side scripting languages, a series of FLOSS systems have held a majority of market share for as long as statistics have been kept (Netcraft 2013; Q-Success 2013). Although Microsoft's Internet Explorer held more than 95 percent of the market for a time, after it squeezed Netscape Navigator out (illegally, according to antitrust adjudications in both the United States and the European Union), the FLOSS browsers Chrome and Firefox now hold nearly half of the market (Vaughan-Nichols 2013). Recent industry surveys suggest that nearly 40 percent of the firms engaged in software development develop and contribute to FLOSS (Lerner and Schankerman 2010). Bonaccorsi and Rossi (2003), von Krogh (2003), and Lorenzi and Rossi (2008) have suggested that FLOSS is more innovative in at least some contexts.

Wikipedia's success, like that of FLOSS, has been connected to its quality, which in turn has been associated with its basis in collaboration from peer production. In a widely cited study, writers at the journal *Nature* presented evidence that a series of Wikipedia articles were comparable in quality to articles on the same subjects in the *Encyclopaedia Britannica* after experts recruited by *Nature* found approximately the same number of errors in each (Giles 2005). Although there is no statistically significant correlation between the number of edits to a Wikipedia article and the number of errors in each article within the *Nature* sample,4 other studies of Wikipedia (Wilkinson and Huberman 2007; Kittur et al. 2009) have shown that articles with more intense collaboration tend to be rated as being of higher quality by other Wikipedia contributors. Similarly, Greenstein and Zhu (2012) have shown that articles on U.S. politics tend to become less biased as the number and intensity of contributors grows. Viégas et al. (2004) and many others have shown that Wikipedia's open editing model makes it practicable to detect and remove vandalism within minutes or even within seconds.

The success of FLOSS and that of Wikipedia have inspired the adoption of peer production in a wide range of industries. Maurer (2010) described instances in which distributed, non-state, non-market action was able to deliver public goods ranging from nanotechnology safety standards to a synthetic DNA anti-terrorism code. Online business models that depend on peer production (and related commons-based approaches) have out-competed those that depend on more traditional price-cleared or firm-centric models of production. For example, Flickr, Photobucket, and Google Images, all of which incorporate peer production into their platform, have produced vast repositories of stock photography and have overshadowed Corbis, the largest stock photography firm using a purely proprietary business model. YouTube, Google video, and Vimeo all rely heavily on peer production for their video content, and each is more popular than the studio-produced models of Hulu, Vevo, or even Netflix (though Netflix, the most widely used among these, is roughly as popular as Vimeo). The peer-produced TripAdvisor is more popular than the travel guides offered by Lonely Planet, Fodor, or Frommer. In restaurant reviews, Yelp is more widely used than alternatives. Both for-profit and non-profit organizations that have incorporated peer-production models have thrived in the networked environment, often overcoming competition from more traditional market-based and firm-based models.

New Directions
Recent research has stepped back from the simple association of peer production with high quality and has problematized earlier celebratory accounts. This has entailed moving beyond the rhetoric of Raymond (1999) to cease taking the high quality of peer-produced resources for granted. Toward that end, this newer wave of scholarship has looked at large data sets of attempts of peer production and has documented that the vast majority of would-be peer-production projects fail to attract communities—an obvious prerequisite to high quality through collaboration. Additional scholarship has suggested that quality cannot be taken

for granted even in situations in which large communities of collaborators are mobilized. Finally, recent work has evaluated peer production in terms of a series of different types and dimensions of quality. In following these three paths, scholars have begun to consider variation within peer-production projects in order to consider when and why peer production leads to different kinds of high-quality outputs.

Contrary to the claims of foundational accounts, studies have suggested that the vast majority of attempts at peer production fail to attract large crowds whose aggregate contributions might invoke Linus's Law. Lacking any substantive collaboration, these projects do not constitute examples of peer production. Healy and Schussman (2003) showed that the median number of contributors to a FLOSS project hosted on Source-Forge is one. Schweik and English (2012) demonstrated that, even among FLOSS projects that have produced successful and sustainable information commons, the median number of contributors is one. A series of studies of remixing suggest that more than 90 percent of projects uploaded to remixing sites are never remixed at all (Luther et al. 2010; Hill and Monroy-Hernández 2013b). Reich et al. (2012) have shown that most attempts to create wikis within classroom environments fail to attract contributions. Hill (2013) has shown that even Wikipedia was preceded by seven other attempts to create freely available, collaborative online encyclopedias—all of which failed to created communities on anything approaching the scale of Wikipedia.

The relative infrequency of peer production within domains such as FLOSS does not necessarily pose a theoretical problem for peer production, since the success of the phenomenon is not defined in terms of the proportion of viable projects. However, the fact that peer production seems so difficult to achieve, and the fact that the vast majority of attempts at peer production never attract a community, are worrisome for those who value the high-quality public goods made available freely on the Internet through peer production. Although this has constituted an inconvenient fact in peer-production practice, it also points to an important opportunity for future research. By focusing only on the projects that successfully mobilize contributors, researchers interested in when peer production occurs or in why it succeeds at producing high-quality outputs have systematically selected on their dependent variables. It will be important for peer-production research to study these failures.

Several critics of peer production have questioned whether large-scale collaborations can ever lead to high-quality outputs. Although relying primarily on anecdotal evidence, Keen (2007) argued that peer production's openness to participation leads to products that are amateurish and of inferior quality. Similarly, Lanier (2010) suggested that Wikipedia's reliance on mass collaboration results in articles that are sterile and lack a single voice. These dismissive treatments are problematic because peer production obviously does lead to high-quality outputs in some situations. A more fruitful approach considers variation in the success of peer production in order to understand when and where it works better and worse. For example, Duguid (2006) raised a series of theoretical challenges for peer production and suggested that it may work less well outside of software development. Benkler (2006) speculated that peer production may be better at producing functional works such as operating-system software and encyclopedias than at producing creative works such as code or art. In support of this idea, research by Benjamin Mako Hill and Andrés Monroy-Hernández (2013a) of the youth-based remixing community Scratch suggested that collaborative works tend to be rated lower than *de novo* projects and that the gap in quality between remixes and non-remixes narrows for code-heavy projects but expands enormously for media-intense works.

Just as peer-produced goods vary in nature and in form, there is enormous variation between and within projects in terms of the dimensions along which quality might be evaluated. For example, scholars have assessed Wikipedia for factual accuracy, scope of coverage, political bias, expert evaluation, and peer evaluation, often drawing different conclusions about the quality of Wikipedia or that of particular articles. Many studies have suggested that Wikipedia has uneven topic coverage. One group of studies has found uneven cross-cultural and cross-language coverage between Wikipedias and shown that the unevenness corresponds to other forms of socioeconomic and cultural inequality (Arazy et al. 2015; Royal and Kapila 2009; Pfeil et al. 2006; Bao et al. 2012; Hecht and Gergle 2009). Inequality in gender participation (see Glott et al. 2010; Hill and Shaw 2013) has also been connected to underproduction of Wikipedia articles on topics related to women by Antin et al. (2011) and Reagle and Rhue (2011). Many explanations for this variation exist, but Halavais and Lackaff (2008) offer a succinct summary: "Wikipedia's topical coverage is driven by the interests of its users, and as a result, the reliability and completeness of Wikipedia is likely to be different depending on the subject area of the article."

A related series of studies have sought to understand the dynamics of collaboration and to understand which features of peer productions support the creation of higher-quality content. This topic has been studied especially closely in the case of Wikipedia, where particular organizational attributes, routines, norms, and technical features affect the quality of individual contributions as well as the final, collaborative product. For example, Priedhorsky et al. (2007) and Halfaker et al. (2009) have provided evidence to support the claim that experienced contributors have a deep sense of ownership and commitment that leads them to make more durable contributions. Other work has shown that persistent subcommunities of contributors can develop stable dynamics of coordination that lead to more productive collaboration and higher-quality products (Kittur and Kraut 2008). However, Half-aker et al. (2011, 2013a) have shown that experienced editors can wield their expertise against new contributors, removing their contributions and undermining their motivation to edit in the future. Loubser (2010) has suggested that this process can shrink the pool from which future experienced contributors and leaders can be drawn. Gorbatai (2012) has demonstrated that harsh treatment of newcomers can also under-mine quality in other, less obvious ways—for example, although expe-rienced editors may perceive low-quality contributions from inexperienced editors as a nuisance, these "newbie" edits also provide a benefit: they attract the attention of experienced community members to improve popular pages they might otherwise have ignored.

An important remaining question is whether, and to what extent, the dynamics, routines, infrastructure, and mechanisms that support peer production in intensively studied projects such as the English-language version of Wikipedia generalize to other communities engaged in other kinds of collaborative activity. Researchers studying peer production may do well to draw on approaches that have been applied to other areas of collective intelligence, such as crowdsourcing, in which more precise analyses of the relationship among tasks, system design, and the quality of collaborative outputs have already been performed (see, e.g., Yu and Nickerson 2011). This is an area in which further comparative study across platforms, communities, contexts, and tasks—research like that done by Roth et al. (2008), Kittur and Kraut (2010), and Schweik and English (2012)—will be fruitful.

Conclusion

In the context of peer production, three of the most important and confounding challenges for scholars of collective intelligence in the social sciences and for legal scholars have concerned organization, motivation, and quality. As we have suggested, these themes do not attempt to be exhaustive, proscriptive, or predictive of future research. In addition to our incomplete reviews of the literature addressing each of these themes, there are many bodies of research that build on, and are related to, peer production that we have not addressed. Nevertheless, our approach illustrates how studies of peer production and collective intelligence have already engaged with sustained debates that range across the social sciences and suggests how they might productively continue to do so. Interested readers seeking a deeper understanding of collective intelligence in the social sciences and in legal scholarship should recognize that this chapter provides an entry point into these topics but is far from a comprehensive summary.

The limitations of earlier scholarship on the subject of peer production underscore the opportunities for continued social scientific study of peer production and collective intelligence from a variety of theoretical and methodological perspectives. We have highlighted several of what we feel are the most promising areas for future work. As we have suggested, studies that engage in comparative analysis, consider failed as well as successful efforts, and identify causal mechanisms can contribute a great deal. Likewise, work that challenges, extends, or elaborates the major theoretical puzzles in any of the three areas will have an impact in to a wide variety of peer-produced phenomena.

Finally, it is possible that evidence from peer production can be used to challenge and extend the literature on collective intelligence, and perhaps to contribute to more general theories of human behavior and cooperation. Peer-production systems combine novel and creative social practices and forms of organization with the creation of truly unprecedented sources of behavioral data. Never, before peer production, had scholars possessed full transcripts of every interaction, communication, and contribution to a collective endeavor or to a particular public good. With the widespread availability of huge data sets from peer-production communities—many of them publicly released under free or open licenses that permit research and reuse—it has become possible to harness the tactics and tools of computational science for the comparative study of social action. Substantively, this means that there are many opportunities to conduct research that cuts across traditional divisions among micro-level, meso-level, and macro-level social analysis. There are also opportunities to revisit classical concerns of social scientific theory through evidence that is granular and exhaustive.

Acknowledgments

Parts of this chapter build upon work published in Y. Benkler, "Peer production and cooperation," in *Handbook on the Economics of the Internet*, ed. J. M. Bauer and M. Latzer (Elgar, 2015).

Notes

1. http://scholar.google.com/scholar?q=intitle%3Awikipedia (accessed July 10, 2013).

2. http://www.alexa.com/siteinfo/wikipedia.org (accessed June 15, 2013).

3. http://opensource.org/about (accessed July 6, 2013).

4. Analysis of data by the authors of this article using public data from nature study published line. http://www.nature.com/nature/journal/v438/n7070/extref/438900a-s1.doc (accessed July 6, 2013).

References

Alexy, O., and M. Leitner. 2011. A fistful of dollars: Are financial rewards a suitable management practice for distributed models of innovation? *European Management Review* 8 (3): 165–185.

Algan, Y., Y. Benkler, M. Fuster Morell, and J. Hergueux. 2013. Cooperation in a peer production economy: Experimental evidence from Wikipedia. Working paper.

Antin, J., and C. Cheshire. 2010. Readers are not free-riders. In *Proceedings of the 2010 ACM Conference on Computer Supported Cooperative Work*. ACM.

Antin, J., C. Cheshire, and O. Nov. 2012. Technology-mediated contributions: Editing behaviors among new Wikipedians. In *Proceedings of the ACM 2012 Conference on Computer Supported Cooperative Work*. ACM.

Antin, J., R. Yee, C. Cheshire, and O. Nov. 2011. Gender differences in Wikipedia editing. In *Proceedings of the 7th International Symposium on Wikis and Open Collaboration*. ACM.

Arazy, O., F. Ortega, O. Nov, L. Yeo, and A. Balila. 2015. Determinants of Wikipedia quality: The roles of global and local contribution inequality. In *Proceedings of the 2015 ACM Conference on Computer Supported Cooperative Work*. ACM.

Bao, P., B. Hecht, S. Carton, M. Quaderi, M. Horn, and D. Gergle. 2012. Omnipedia: Bridging the Wikipedia language gap. In *Proceedings of the 2012 ACM Annual Conference on Human Factors in Computing Systems*. ACM.

Beenen, G., K. Ling, X. Wang, K. Chang, D. Frankowski, P. Resnick, and R. E. Kraut. 2004. Using social psychology to motivate contributions to online communities. In *Proceedings of the 2004 ACM Conference on Computer Supported Cooperative Work*. ACM.

Belenzon, S., and M. A. Schankerman. 2008. *Motivation and sorting in open source software innovation. Scholarly Paper 1401776.* Social Science Research Network.

Benkler, Y. 2001. The battle over the institutional ecosystem in the digital environment. *Communications of the ACM* 44 (2): 84–90.

Benkler, Y. 2002. Coase's penguin, or, Linux and the nature of the firm. *Yale Law Journal* 112 (3): 369–446.

Benkler, Y. 2004. Sharing nicely: On shareable goods and the emergence of sharing as a modality of economic production. *Yale Law Journal* 114 (2): 273–358.

Benkler, Y. 2006. *The Wealth of Networks: How Social Production Transforms Markets and Freedom*. Yale University Press.

Benkler, Y. 2009. Law, policy, and cooperation. In *Government and Markets: Toward A New Theory of Regulation*, ed. E. Balleisen and D. Moss. Cambridge University Press.

Benkler, Y. 2011. *The Penguin and the Leviathan: How Cooperation Triumphs over Self-Interest*. Crown Business.

Benkler, Y. 2013. Peer production and cooperation. In *Handbook on the Economics of the Internet*, ed. J. M. Bauer and M. Latzer. Elgar.

Benkler, Y., and H. Nissenbaum. 2006. Commons-based peer production and virtue. *Journal of Political Philosophy* 14 (4): 394–419.

Bonaccorsi, A., and C. Rossi. 2003. Why open source software can succeed. *Research Policy* 32 (7): 1243–1258.

Bowles, S., and S.-H. Hwang. 2008. Social preferences and public economics: Mechanism design when social preferences depend on incentives. *Journal of Public Economics* 92 (8–9): 1811–1820.

Bowles, S., and S. Polania-Reyes. 2012. Economic incentives and social preferences: Substitutes or complements? *Journal of Economic Literature* 50 (2): 368–425.

Brabham, D. C. 2013. *Crowdsourcing*. MIT Press.

Bryant, S. L., A. Forte, and A. Bruckman. 2005. Becoming Wikipedian: Transformation of participation in a collaborative online encyclopedia. In *Proceedings of the 2005 international ACM SIGGROUP Conference on Supporting Group Work*. ACM.

Burke, M., and R. Kraut. 2008. Mopping up: Modeling Wikipedia promotion decisions. In *Proceedings of the ACM 2008 Conference on Computer Supported Cooperative Work*. ACM.

Butler, B. S., E. Joyce, and J. Pike. 2008. Don't look now, but we've created a bureaucracy: The nature and roles of policies and rules in Wikipedia. In *Proceedings of the SIGCHI Conference on Human Factors in Computing Systems*. ACM.

Cheshire, C. 2007. Selective incentives and generalized information exchange. *Social Psychology Quarterly* 70: 82–100.

Cheshire, C., and J. Antin. 2008. The social psychological effects of feedback on the production of internet information pools. *Journal of Computer-Mediated Communication* 13 (3): 705–727.

Coleman, E. G. 2012. *Coding Freedom: The Ethics and Aesthetics of Hacking*. Princeton University Press.

Coleman, G., and B. M. Hill. 2004. The social production of ethics in Debian and free software communities: Anthropological lessons for vocational ethics. In *Free/Open Source Software Development*, ed. S. Koch. Idea Group.

Crowston, K., K. Wei, J. Howison, and A. Wiggins. 2010. Free/libre open source software: What we know and what we do not know. *ACM Computing Surveys* 44 (2): 1–35.

David, P. A., and J. S. Shapiro. 2008. Community-based production of open-source software: What do we know about the developers who participate? *Information Economics and Policy* 20 (4): 364–398.

Duguid, P. 2006. Limits of self-organization: Peer production and "laws of quality." *First Monday* 11 (10).

Fisher, D., M. Smith, and H. T. Welser. 2006. You are who you talk to: Detecting roles in Usenet newsgroups. In *Proceedings of the 39th Annual Hawaii International Conference on System Sciences*, volume 3. IEEE Computer Society.

Forte, A., and A. Bruckman. 2008. Scaling consensus: Increasing decentralization in Wikipedia governance. In *Proceedings of the 41st Annual Hawaii International Conference on System Science*. IEEE.

Forte, A., V. Larco, and A. Bruckman. 2009. Decentralization in Wikipedia governance. *Journal of Management Information Systems* 26 (1): 49–72.

Frey, B. S. 1997. A constitution for knaves crowds out civic virtues. *Economic Journal* 107 (443): 1043–1053.

Frey, B. S., and R. Jegen. 2001. Motivation crowding theory. *Journal of Economic Surveys* 15 (5): 589–611.

Fuster Morell, M. 2010. Governance of Online Creation Communities: Provision of Infrastructure for the Building of Digital Commons. PhD dissertation, European University Institute, Florence.

Ghosh, R. A. 1998. Cooking pot markets: An economic model for the trade in free goods and services on the internet. *First Monday* 3 (3).

Ghosh, R. A. 2005. Understanding free software developers: Understandings from the FLOSS study. In *Perspectives on Free and Open Source Software*, ed. J. Feller et al. MIT Press.

Ghosh, R. A., R. Glott, B. Krieger, and G. Robles. 2002. *Free/Libre and Open Source Software: Survey and Study*. Technical report, Maastricht Economic Research Institute on Innovation and Technology. University of Maastricht.

Ghosh, R. A., and V. V. Prakash. 2000. The Orbiten free software survey. *First Monday* 5 (7).

Giles, J. 2005. Internet encyclopaedias go head to head. *Nature* 438 (7070): 900–901.

Glott, R., R. Ghosh, and P. Schmidt. 2010. *Wikipedia survey*. Technical report. UNU-MERIT.

Gorbatai, A. 2012. Aligning collective production with demand: Evidence from Wikipedia. Working paper.

Greenstein, S., and F. Zhu. 2012. Is Wikipedia biased? *American Economic Review* 102 (3): 343–348.

Halavais, A., and D. Lackaff. 2008. An analysis of topical coverage of Wikipedia. *Journal of Computer-Mediated Communication* 13 (2): 429–440.

Halfaker, A., R. S. Geiger, J. T. Morgan, and J. Riedl. 2013a. The rise and decline of an open collaboration system how Wikipedia's reaction to popularity is causing its decline. *American Behavioral Scientist* 57 (5): 664–688.

Halfaker, A., O. Keyes, and D. Taraborelli. 2013b. Making peripheral participation legitimate: Reader engagement experiments in Wikipedia. In *Proceedings of the 2013 Conference on Computer Supported Cooperative Work*. ACM.

Halfaker, A., A. Kittur, R. Kraut, and J. Riedl. 2009. A jury of your peers: Quality, experience and ownership in Wikipedia. In *Proceedings of the 5th International Symposium on Wikis and Open Collaboration*. ACM.

Halfaker, A., A. Kittur, and J. Riedl. 2011. Don't bite the newbies: How reverts affect the quantity and quality of Wikipedia work. In *Proceedings of the 7th International Symposium on Wikis and Open Collaboration*. ACM.

Hars, A., and S. Ou. 2002. Working for free? motivations for participating in open-source projects. *International Journal of Electronic Commerce* 6 (3): 25–39.

Healy, K., and A. Schussman. 2003. The ecology of open-source software development. Working paper.

Hecht, B., and D. Gergle. 2009. Measuring self-focus bias in community-maintained knowledge repositories. In *Proceedings of the Fourth International Conference on Communities and Technologies*. ACM.

Hill, B. M. 2011. Causal effects of a reputation-based incentive in an peer production community. Working paper.

Hill, B. M. 2013. Essays on Volunteer Mobilization in Peer Production. PhD dissertation, Massachusetts Institute of Technology.

Hill, B. M., and A. Monroy-Hernández. 2013a. The cost of collaboration for code and art: Evidence from a remixing community. In *Proceedings of the 2013 Conference on Computer Supported Cooperative Work*. ACM.

Hill, B. M., and A. Monroy-Hernández. 2013b. The remixing dilemma the trade-off between generativity and originality. *American Behavioral Scientist* 57 (5): 643–663.

Hill, B. M., and A. Shaw. 2013. The Wikipedia gender gap revisited: Characterizing survey response bias with propensity score estimation. *PLoS ONE* 8 (6): e65782.

Hill, B. M., A. Shaw, and Y. Benkler. 2012. Status, social signalling and collective action in a peer production community. Working paper.

Hindman, M. 2008. *The Myth of Digital Democracy*. Princeton University Press.

Howison, J., M. Conklin, and K. Crowston. 2006. FLOSSmole. *International Journal of Information Technology and Web Engineering* 1 (3): 17–26.

Jemielniak, D. 2014. *Common Knowledge? An Ethnography of Wikipedia*. Stanford University Press.

Keegan, B., and D. Gergle. 2010. Egalitarians at the gate. In *Proceedings of the 2010 ACM Conference on Computer Supported Cooperative Work*. ACM.

Keen, A. 2007. *The Cult of the Amateur: How Today's Internet Is Killing Our Culture*. Crown Business.

Kelty, C. 2008. *Two Bits: The Cultural Significance of Free Software.* Duke University Press.

King, A. A., and K. R. Lakhani. 2011. The contingent effect of absorptive capacity: An open innovation analysis. Working paper 11-102, Harvard Business School.

Kittur, A., and R. E. Kraut. 2008. Harnessing the wisdom of crowds in Wikipedia: Quality through coordination. In *Proceedings of the 2008 ACM Conference on Computer Supported Cooperative Work.* ACM.

Kittur, A., and R. E. Kraut. 2010. Beyond Wikipedia: Coordination and conflict in online production groups. In *Proceedings of the 2010 ACM Conference on Computer Supported Cooperative Work.* ACM.

Kittur, A., B. Lee, and R. E. Kraut. 2009. Coordination in collective intelligence: The role of team structure and task interdependence. In *Proceedings of the 27th International Conference on Human Factors in Computing Systems.* ACM.

Kollock, P. 1999. The economies of online cooperation: Gifts and public goods in cyberspace. In *Communities in Cyberspace,* ed. M. Smith and P. Kollock. Routledge.

Konieczny, P. 2009. Governance, organization, and democracy on the internet: The iron law and the evolution of Wikipedia. *Sociological Forum* 24 (1): 162–192.

Kraut, R. E., and P. Resnick. 2012. *Building Successful Online Communities: Evidence-Based Social Design.* MIT Press.

Kreiss, D., M. Finn, and F. Turner. 2011. The limits of peer production: Some reminders from Max Weber for the network society. *New Media & Society* 13 (2): 243–259.

Kriplean, T., I. Beschastnikh, and D. W. McDonald. 2008. Articulations of wikiwork: Uncovering valued work in Wikipedia through barnstars. In *Proceedings of the 2008 ACM Conference on Computer Supported Cooperative Work.* ACM.

Lakhani, K. 2006. The Core and the Periphery in Distributed and Self-Organizing Innovation Systems. PhD dissertation, Sloan School of Management, Massachusetts Institute of Technology.

Lakhani, K., and B. Wolf. 2005. Why hackers do what they do: Understanding motivation and effort in free/open source software projects. In *Perspectives on Free and Open Source Software,* ed. J. Feller et al. MIT Press.

Lampe, C., and P. Resnick. 2004. Slash(dot) and burn: Distributed moderation in a large online conversation space. In *Conference on Human Factors in Computing Systems.* ACM.

Lampe, C., R. Wash, A. Velasquez, and E. Ozkaya. 2010. Motivations to participate in online communities. In *Proceedings of the 28th International Conference on Human Factors in Computing Systems.* ACM.

Lampe, C. A., E. Johnston, and P. Resnick. 2007. Follow the reader: Filtering comments on Slashdot. In *Proceedings of the SIGCHI Conference on Human Factors in Computing Systems.* ACM.

Lanier, J. 2010. *You Are Not a Gadget: A Manifesto.* Knopf.

Lerner, J., and M. Schankerman. 2010. *The Comingled Code: Open Source and Economic Development.* MIT Press.

Lerner, J., and J. Tirole. 2002. Some simple economics of open source. *Journal of Industrial Economics* 50 (2): 197–234.

Leskovec, J., D. Huttenlocher, and J. Kleinberg. 2010. Governance in social media: A case study of the Wikipedia promotion process. Presented at International AAAI Conference on Weblogs and Social Media, Washington.

Lorenzi, D., and C. Rossi. 2008. Assessing innovation in the software sector: Proprietary vs. FOSS production mode. preliminary evidence from the Italian case. In *Open Source Development, Communities and Quality,,* ed. B. Russo, E. Damiani, S. Hissam, B. Lundell, and G. Succi. Springer.

Loubser, M. 2010. Organisational Mechanisms in Peer Production: The Case of Wikipedia. DPhil dissertation, Oxford Internet Institute, Oxford University.

Luther, K., K. Caine, K. Zeigler, and A. Bruckman. 2010. Why it works (when it works): Success factors in online creative collaboration. In *Proceedings of the ACM Conference on Supporting Group Work.* ACM.

Luther, K., C. Fiesler, and A. Bruckman. 2013. Redistributing leadership in online creative collaboration. In *Proceedings of the 2013 Conference on Computer Supported Cooperative Work.* ACM.

Maurer, S. 2010. *Five easy pieces: Case studies of entrepreneurs who organized private communities for a public purpose. Scholarly paper 1713329.* Social Science Research Network.

McDonald, D. W., S. Javanmardi, and M. Zachry. 2011. Finding patterns in behavioral observations by automatically labeling forms of wikiwork in barnstars. In *Proceedings of the 7th International Symposium on Wikis and Open Collaboration.* ACM.

Moglen, E. 1999. Anarchism triumphant: Free software and the death of copyright. *First Monday* 4 (8).

Netcraft. 2013. January 2013 Web server survey. http://news.netcraft.com/archives/2013/01/07/january-2013-web-server-survey-2.html.

Nov, O. 2007. What motivates Wikipedians? *Communications of the ACM* 50 (11): 60–64.

O'Mahony, S., and B. A. Bechky. 2008. Boundary organizations: Enabling collaboration among unexpected allies. *Administrative Science Quarterly* 53 (3): 422–459.

O'Mahony, S., and F. Ferraro. 2007. The emergence of governance in an open source community. *Academy of Management Journal* 50 (5): 1079–1106.

Ortega, F. 2009. *Wikipedia: A Quantitative Analysis.* PhD dissertation, Universidad Rey Juan Carlos, Madrid.

Osterloh, M., J. Frost, and B. S. Frey. 2002. The dynamics of motivation in new organizational forms. *International Journal of the Economics of Business* 9 (1): 61–77.

Ostrom, E. 1990. *Governing the Commons: The Evolution of Institutions for Collective Action.* Cambridge University Press.

Panciera, K., A. Halfaker, and L. Terveen. 2009. Wikipedians are born, not made: A study of power editors on Wikipedia. In Proceedings of the ACM 2009 International Conference on Supporting Group Work. ACM.

Pentzold, C. 2011. Imagining the Wikipedia community: What do Wikipedia authors mean when they write about their "community"? *New Media & Society* 13 (5): 704–721.

Pfeil, U., P. Zaphiris, and C. S. Ang. 2006. Cultural differences in collaborative authoring of Wikipedia. *Journal of Computer-Mediated Communication* 12 (1): 88–113.

202 Law, Communications, Sociology, Political Science, and Anthropology

Priedhorsky, R., J. Chen, S. T. K. Lam, K. Panciera, L. Terveen, and J. Riedl. 2007. Creating, destroying, and restoring value in Wikipedia. In Proceedings of the 2007 International ACM *Conference* on Supporting Group Work. ACM.

Q-Success 2013. Usage statistics and market share of server-side programming languages for websites, July.

Rafaeli, S., and Y. Ariel. 2008. Online motivational factors: Incentives for participation and contribution in Wikipedia. In *Psychological Aspects of Cyberspace: Theory, Research, Applications*, ed. A. Barak. Cambridge University Press.

Ransbotham, S., and G. C. Kane. 2011. Membership turnover and collaboration success in online communities: Explaining rises and falls from grace in Wikipedia. *MIS Quarterly* 35 (3): 613–627.

Raymond, E. S. 1999. *The Cathedral and the Bazaar: Musings on Linux and Open Source by an Accidental Revolutionary*. O'Reilly.

Reagle, J. 2010. *Good Faith Collaboration: The Culture of Wikipedia*. MIT Press.

Reagle, J., and L. Rhue. 2011. Gender bias in Wikipedia and Britannica. *International Journal of Communication* 5.

Reich, J., R. Murnane, and J. Willett. 2012. The state of wiki usage in U.S. K–12 schools: Leveraging Web 2.0 data warehouses to assess quality and equity in online learning environments. *Educational Researcher* 41 (1): 7–15.

Restivo, M., and A. van de Rijt. 2012. Experimental study of informal rewards in peer production. *PLoS ONE* 7 (3): e34358.

Restivo, M., and A. van de Rijt. 2014. No praise without effort: Experimental evidence on how rewards affect Wikipedia's contributor community. *Information Communication and Society* 17 (4): 451–462.

Roth, C., D. Taraborelli, and N. Gilbert. 2008. Measuring wiki viability: An empirical assessment of the social dynamics of a large sample of wikis. In *Proceedings of the 4th International Symposium on Wikis*. ACM.

Royal, C., and D. Kapila. 2009. What's on Wikipedia, and what's not …? Assessing completeness of information. *Social Science Computer Review* 27 (1): 138–148.

Savitz, E. 2012. The Forbes Fast Tech 25: Our annual list of growth kings. www.forbes.com, May 21.

Schweik, C. M., and R. C. English. 2012. *Internet Success: A Study of Open-Source Software Commons*. MIT Press.

Shah, S. K. 2006. Motivation, governance, and the viability of hybrid forms in open source software development. *Management Science* 52 (7): 1000–1014.

Shaw, A. 2012. Centralized and decentralized gatekeeping in an open online collective. *Politics & Society* 40 (3): 349–388.

Shaw, A., and B. M. Hill. 2014. Laboratories of oligarchy? How the iron law applies to peer production. *Journal of Communication* 64 (2): 215–238.

Spaeth, S., M. Stuermer, and G. V. Krogh. 2010. Enabling knowledge creation through outsiders: Towards a push model of open innovation. *International Journal of Technology Management* 52 (3/4): 411–431.

Vaughan-Nichols, S. J. 2013. The Web browser wars continue, and #1 is… well, that depends on whom you ask. http://www.zdnet.com/article/the-web-browser-wars -continue-and-1-is-well-that-depends-on-whom-you-ask/.

Viégas, F. B., M. Wattenberg, and K. Dave. 2004. Studying cooperation and conflict between authors with history flow visualizations. In *Proceedings of the SIGCHI Conference on Human Factors in Computing Systems*. ACM.

Viégas, F. B., M. Wattenberg, M. M. McKeon, and D. Schuler. 2007. The hidden order of Wikipedia. In *Online Communities and Social Computing*. Springer.

von Hippel, E., and G. von Krogh. 2003. Open source software and the "private-collective" innovation model: Issues for organization science. *Organization Science* 14 (2): 209–223.

von Krogh, G. 2003. Open-source software development. *MIT Sloan Management Review*, spring: 14–18.

von Krogh, G., S. Haefliger, S. Spaeth, and M. W. Wallin. 2012. Carrots and rainbows: Motivation and social practice in open source software development. *Management Information Systems Quarterly* 36 (2): 649–676.

von Krogh, G., S. Spaeth, and K. R. Lakhani. 2003. Community, joining, and specialization in open source software innovation: A case study. *Research Policy* 32 (7): 1217–1241.

von Krogh, G., and E. von Hippel. 2006. The promise of research on open source software. *Management Science* 52 (7): 975–983.

Wang, L. S., J. Chen, Y. Ren, and J. Riedl. 2012. Searching for the goldilocks zone: Trade-offs in managing online volunteer groups. In *Proceedings of the ACM 2012 Conference on Computer Supported Cooperative Work*. ACM.

Weber, S. 2004. *The Success of Open Source*. Harvard University Press.

Welser, H. T., D. Cosley, G. Kossinets, A. Lin, F. Dokshin, G. Gay, and M. Smith. 2011. Finding social roles in Wikipedia. In *Proceedings of the 2011 iConference*. ACM.

West, J., and S. O'Mahony. 2005. Contrasting community building in sponsored and community founded open source projects. In *Proceedings of the 38th Annual Hawaii International Conference on System Sciences*. IEEE Computer Society.

Wilkinson, D. M., and B. A. Huberman. 2007. Cooperation and quality in Wikipedia. In *Proceedings of the 2007 International Symposium on Wikis*. ACM.

Willer, R. 2009. Groups reward individual sacrifice: The status solution to the collective action problem. *American Sociological Review* 74 (1): 23–43.

Williamson, O. E. 1985. *The Economic Institutions of Capitalism*. Free Press.

Wu, F., D. M. Wilkinson, and B. A. Huberman. 2009. Feedback loops of attention in peer production. In *Proceedings of the 2009 International Conference on Computational Science and Engineering*, volume 4. IEEE Computer Society.

Yasseri, T., R. Sumi, A. Rung, A. Kornai, and J. Kertesz. 2012. Dynamics of conflicts in Wikipedia. *PLoS ONE* 7 (6): e38869.

Yu, L., and J. V. Nickerson. 2011. Cooks or cobblers? In *Proceedings of the SIGCHI Conference on Human Factors in Computing System*. ACM.

Zhang, X. M., and F. Zhu. 2011. Group size and incentives to contribute: A natural experiment at Chinese Wikipedia. *American Economic Review* 101: 1601–1615.

Zhu, H., R. Kraut, and A. Kittur. 2012. Organizing without formal organization: Group identification, goal setting and social modeling in directing online production. In *Proceedings of the ACM 2012 Conference on Computer Supported Cooperative Work*. ACM.

Zhu, H., R. E. Kraut, Y.-C. Wang, and A. Kittur. 2011. Identifying shared leadership in Wikipedia. In *Proceedings of the 2011 Annual Conference on Human Factors in Computing Systems*. ACM.

Conclusion

Thomas W. Malone

In the preceding chapters, we have seen how many different disciplines can contribute to our understanding of collective intelligence. To summarize how all these different perspectives relate to each other, let us consider some of the key elements of any collectively intelligent system. As figure C.1 suggests, we can analyze any collectively intelligent system using four questions: What is being done? Who is doing it? Why are they doing it? How are they doing it?

What Is Being Done?

In systems involving humans, as was discussed in the chapter on human–computer interaction, the chapter on organizational behavior, and the

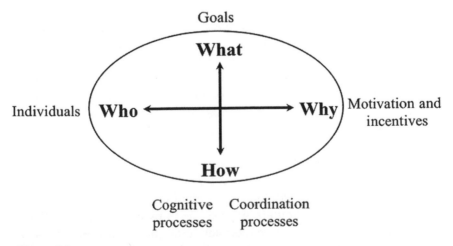

Figure C.1
Elements of collectively intelligent systems. Adapted from T. Malone, R. Laubacher, and C. Dellarocas, "The collective intelligence genome," *MIT Sloan Management Review* 51, no. 3 (2010): 21–31 and J. R. Galbraith, *Designing Organizations* (Jossey-Bass, 2002).

chapter on law and other disciplines, groups often explicitly choose their goals. In these cases, determining how goals are selected is critical for understanding or designing the system.

But, as was suggested in the introduction and in the chapter on economics, it is also often useful to evaluate systems with respect to goals that may not be held by the individuals involved in the group. For instance, even if no participant in a market has the goal of efficiently allocating societal resources, the chapter on economics describes how this often happens anyway. And as the chapter on biology suggests, in systems (e.g., ant colonies) where the individuals are not humans, often the only way we can evaluate the collective intelligence of the system is with respect to "goals" we—as observers—attribute to the system. In these cases, picking useful goals with respect to which to evaluate the system is critical to making research progress.

Who Is Doing It?

For a group to act intelligently, the individuals in it must somehow—together—have the characteristics needed to perform the necessary activities. Different disciplines have studied this issue in different ways. At a simple level, for instance, as we saw in the chapter on biology, the functioning of an ant colony depends on having different individuals—such as queens and workers—with different, but complementary, capabilities.

More subtly, as we saw in the introduction to the chapter on economics and in the chapters on cognitive psychology and organizational behavior, the ability of a group to perform a task well may depend on the diversity of the group members. For instance, groups estimating the number of jellybeans in a jar do not perform well if their members have the same biases (cognitive psychology chapter). But if group members are *too* diverse in cognitive style, their performance may suffer then, too (organizational behavior chapter).

Why Are They Doing It?

In some cases, such as those of simple computer systems and groups of non-human animals, understanding the motivations and incentives of the individual actors can be relatively straightforward. But in most systems involving people, understanding how people's motivations and incentives determine their actions is both non-trivial and essential to understanding the system.

The discipline of economics, for instance, has devoted much of its attention to analyzing how incentives affect the outcomes of group interactions in markets, organizations, and many other situations (economics chapter). But humans are also often motivated by a wide range of factors that go far beyond simple financial gain.

The chapters on human–computer interaction, organizational behavior, and law and other disciplines all talk about the importance of both extrinsic and intrinsic motivations. Contributors to Wikipedia and open-source software projects, for instance, are usually motivated by intrinsic motivations such as desires for intellectual stimulation, recognition, and the pure pleasure of doing the work (chapter on law and other disciplines). And in some cases, providing extrinsic incentives (e.g., money) can actually "crowd out" intrinsic motivations for doing creative or other tasks (organizational behavior chapter).

How Are They Doing It?

Understanding what the individuals are doing to achieve the goals is crucial to understanding any collectively intelligent system. One of the most useful ways of doing this is in terms of the basic *cognitive processes* necessary for intelligence and the basic *coordination processes* necessary for individuals to work together.

Cognitive processes

Cognitive psychology has identified a set of basic cognitive processes needed for any kind of intelligence—collective or otherwise. These processes include the following:

Decision making The chapters on economics, biology, cognitive psychology, and organizational behavior all discussed how groups of individuals make decisions. These group decisions ranged from allocating financial resources in an economy to aggregating human probability estimates for future events to deciding how many ants would look for food in various locations. We saw, for instance, that there are some surprising similarities between economic decision making in markets and evolutionary decision making in biology. And we saw how many of the kinds of decisions that are often made in organizational hierarchies can also be made in much more decentralized ways like those used in markets and groups of animals. For example, in the chapter on biology we saw how groups of ants—each with much less individual intelligence than a human has—can make collective choices (about such things as where and when to look for food) using decision processes that are much more decentralized and local than the decision processes that human groups usually use.

Problem solving The chapters on human–computer interaction, artificial intelligence, organizational behavior, and law and other disciplines all focused on how groups of people can do more than just *decide* among a small number of pre-specified options, but how they can also *create* things that solve problems or achieve other goals. For instance, these chapters discussed how groups of people and computers create encyclopedia articles, develop computer software, solve mathematics problems, play chess, and recognize objects in pictures. Even though much of this work has been done in different research communities, it all deals with similar issues such as how to *generate* possible solutions and how to *evaluate* them.

Perceiving In order to make their decisions sensibly, the groups described in the chapters that discussed decision making (those on economics, biology, cognitive psychology, and organizational behavior) also need to perceive their environment. For instance, one reason the decentralized decision-making methods in economic markets and ant colonies often work well is that, in both cases, the individuals that participate in decision making are also acting as sensors that bring together their perceptions from a wide range of different situations. Perhaps lessons from cognitive psychology about individual perception can help understand these kinds of collective perception, too.

Remembering As the chapter on organizational behavior suggests, groups remember things differently than individuals do. For instance, a "group mind" can remember things, even when most individuals in the group don't. What's needed is for different people to specialize in remembering different things, and for everyone to know how to find who knows specific things. While most work on this kind of transactive memory has focused on groups of people, the possibilities for increasing the collective intelligence of a group by sharing this memory load with computer systems seem immense.

Learning By learning, here, we don't mean just remembering things; we mean improving performance over time. For instance, the chapter on organizational behavior described how traditional human organizations improve their performance by changing their processes or learning which of their members are good at which tasks. The chapter on artificial intelligence described how artificially intelligent systems can do similar things. They can, for example, automatically determine which workers in a crowd are best at which tasks, which workflows work best, and what parameters of a given workflow will optimize its performance. The chapter describes how these techniques have been used for crowdsourced tasks, but an intriguing possibility is that they may, in the future, be increasingly applied in more traditional organizations, too.

Coordination Processes

When *groups* of individuals perform these cognitive processes collectively, the individuals have to do more than just perform the basic activities themselves; they also have to somehow coordinate their activities. As was described in the chapters on economics, organizational behavior, and law and other disciplines, this includes questions like the following: How is the overall problem divided into sub-parts? How are sub-parts assigned to different individuals? How are the activities of individuals working on the sub-parts combined and coordinated?

In the chapter on organizational behavior, for example, we saw various ways of subdividing work in hierarchical organizations (e.g., into functional departments or product divisions), and various ways that managers in these organizations coordinate the work of their own and others' subordinates. But we saw in the economics chapter that decisions about how to subdivide, allocate, and coordinate work can also be made by market interactions among competing individuals and firms.

And in the chapters on human–computer interaction, artificial intelligence, and law and other disciplines, we saw examples of new ways of dividing work into much smaller pieces (such as "microtasks") and new ways of allocating work to people and computers. In some cases, such as that of Wikipedia, this involves machines that "merely" connect intelligent humans so they can work together more effectively. In other cases, such as that of the Google search algorithm, this involves machines doing "intelligent" actions that interpret signals from and complement what humans can do.

Concluding Thoughts

We believe that there are many opportunities for interdisciplinary research on these issues. In many cases, researchers in one discipline may be able to directly use results that are already well known in other disciplines. For instance, in the chapter on cognitive psychology and that on organizational behavior we saw how theories of individual learning, decision making, and intelligence have stimulated research on how these things happen in groups and organizations. How many more such opportunities for interdisciplinary cross-fertilization are still waiting to be discovered?

One promising way to facilitate such work is by developing better theories about generalizable patterns or motifs for each of the different elements of collective intelligence described above. For instance, the chapter on biology describes a pattern that is used to regulate the volume of activity in both the foraging behavior of ants and the transmission of information on the Internet. Perhaps, by recognizing, combining, and modifying patterns that already exist, we will be able to invent new patterns for collective intelligence that have never even been tried before.

We believe that the above-mentioned disciplines can work together to develop a much richer understanding of how collective intelligence works and what it makes possible. Our goal in this book has been to encourage just this sort of interdisciplinary cross-fertilization. We hope that this, in turn, will help us better understand the kinds of collective intelligence that have existed for millennia. And, perhaps more important, we hope it will help us create new kinds of collective intelligence that have never existed on our planet before.

Contributors

Eytan Adar

University of Michigan

Ishani Aggarwal

Georgia Institute of Technology

Yochai Benkler

Harvard University

Michael S. Bernstein

Stanford University

Jeffrey P. Bigham

Carnegie Mellon University

Jonathan Bragg

University of Washington

Deborah M. Gordon

Stanford University

Benjamin Mako Hill

University of Washington

Christopher H. Lin

University of Washington

Andrew W. Lo

Massachusetts Institute of Technology

Thomas W. Malone

Massachusetts Institute of Technology

Mausam

Indian Institute of Technology, Delhi

Brent Miller

University of California, Irvine

Aaron Shaw

Northwestern University

Mark Steyvers

University of California, Irvine

Daniel S. Weld

University of Washington

Anita Williams Woolley

Carnegie Mellon University

Index